THE ROYAL HORTICULTURAL SOCIETY
PRACTICAL GUIDES

DROUGHT-
RESISTANT
GARDENING

THE ROYAL HORTICULTURAL SOCIETY
PRACTICAL GUIDES

DROUGHT-RESISTANT
GARDENING

PETER ROBINSON

DORLING KINDERSLEY

LONDON • NEW YORK • SYDNEY • MOSCOW

www.dk.com

A DORLING KINDERSLEY BOOK
www.dk.com

SERIES EDITOR Pamela Brown
SERIES ART EDITOR Stephen Josland
ART EDITOR Rachael Parfitt

MANAGING EDITOR Louise Abbott
MANAGING ART EDITOR Lee Griffiths

DTP DESIGNER Matthew Greenfield

PRODUCTION Ruth Charlton, Mandy Inness

First published in Great Britain in 1999
by Dorling Kindersley Limited,
9 Henrietta Street, London WC2E 8PS

Visit us on the World Wide Web at http://www.dk.com

Copyright © 1999 Dorling Kindersley Limited, London

A CIP catalogue for this book is available from the British Library.
ISBN 0 7513 06975

Reproduced by Colourscan, Singapore
Printed and bound by Star Standard Industries, Singapore

CONTENTS

SAVING WATER IN THE GARDEN

LOOKING TO THE FUTURE

Each year, rainfall records around the world emphasize just how unpredictable weather can be. What is certain, though, is that the daily, ever-rising demand for water means that gardeners everywhere need to think carefully about how they use – and sometimes waste – this precious and limited resource and how, with a little imagination, they can easily help to conserve it.

A CHANGE OF DIRECTION

Drought is relative – meaning different things in different parts the world. But almost everywhere, water now has its price, both economically and, more importantly, in terms of the environment. Even in temperate regions, gardeners are facing prolonged shortages and can no longer simply turn on the hose. For some, saving water may only involve introducing ways of storing rainwater and improving moisture conservation in the soil. For others, it is providing the impetus for change to a more inventive approach, planting only drought-tolerant plants or even looking for inspiration to the dry areas of the globe where ornament and structure play a key role. All these ideas are helping pave the way for more ecologically conscious gardens in the 21st century.

BLUE AND GREEN *A mosaic pool and brightly painted walls are the dominant features in this tiny courtyard garden. A cleverly placed mirror makes the garden appear larger than it really is, as well as making the planting look more luxuriant.*

WHERE THE WIND BLOWS *Grasses, broom and anthemis stand up to wind and a lack of water.*

Old Style, New Methods

Even in the most conventional garden, simple changes in cultivation techniques and planting ideas can be extremely effective without drastically affecting style. Mulches (*see p.22*) help to limit evaporation from the soil, water butts or tanks (*see p.45*) can be used to store rainwater from times of plenty and, if need be, domestic waste water can be recycled. Small shade trees help to provide cool pockets of air, and windbreaks (*see p.15*) greatly reduce the drying effect of wind. When combined with plants that suit dry conditions (*see pp.29–37*), using only a few of these methods will cut down on the need for supplementary watering. On the other hand, this may be the time to take a more fundamental look at garden design.

Mediterranean Effect

In Mediterranean countries, it was realized long ago that even in a hot, dry climate, gardens could offer shade and refreshment yet use a minimum of water. These small walled gardens and courtyards, often of Moorish influence, have had an enormous impact on garden design, particularly in countries sharing a similar climate. The emphasis is on minimal planting, usually in conjunction with a water feature such as a fountain or canal, with great importance placed on the mix of colours and textures

> For the supreme design source, look at what works in the wild

of non-plant materials. This is a style that has become popular in cooler climates, too, especially in small town gardens, where it is possible to recreate an enclosed atmosphere. Wall fountains, raised pools and small canals that rely on small submersible pumps for re-circulation use very little water.

SWATHES OF COLOUR
Yellow achilleas, red lychnis and red hot pokers (Kniphofia) *combine in a planting that, once established, needs no additional watering.*

◀ BARE MINIMUM
In this Australian garden, the minimalist approach has been taken to the extreme, with cacti alone forming the planting. It would be easy to introduce a water feature and shading for the seating.

▼ ON THE BEACH
In his inimitable garden in the shingle of the Kent coast, artist Derek Jarman mixed seashore flotsam with British native species and Californian poppies (Eschscholzia).

PRAIRIE THEME

In complete contrast to the courtyards of the Mediterranean are the informal "naturalistic" gardens inspired by the dry expanses and prairies of the larger continents. This type of landscape is often dominated by wild grasses and scrub, punctuated by scatterings of wild flowers. When copied in gardens, the effect can be extremely successful. Plants will need watering until they are settled in, but thereafter this type of planting places little dependence on extra water. Under a less brilliant sun, the colours have a pleasing harmony in both winter and summer and, in wind, the light, often feathery flowers have a soothing quality. With this kind of style, it is not essential to look too far from home for planting ideas. You can use drought-tolerant native species in a scheme that blends with local landscapes. Or you may have the best of both worlds, creating just this type of effect (*see pp.18 and 30*) but including plants that have been bred for improved flower colour, length of season or sometimes resistance to disease.

ASSESSING THE PROBLEMS

WHEN STARTING A NEW GARDEN, or redesigning or deciding the best way to reduce water consumption in an existing garden, it is important to assess just how much of a problem you face. Many factors affect the rate at which a garden may lose or use water, and that in turn may vary in different parts of the same site. Soil type, wind, sun and shade all play their part, as well as the frequency with which the rain clouds gather.

THE RAIN FACTOR

The amount of rain a garden receives is inevitably the major factor influencing the need to conserve water and your choice of design and plants. Different regions receiving the same total annual rainfall may face very different problems. The effect of small regular amounts will not be the same as that of infrequent but heavy storms or the majority of the rain falling in one season, and plants have developed a whole variety of adaptations to suit conditions. Where annual rainfall is generally below 45cm, lawns will wither without irrigation and the plants should be limited to those that can resist drought (*see pp.32–35*). An inexpensive gauge will record just how much rain any garden receives (and also if the amount varies within the same garden, due to tree cover or rainshadow).

LUXURY LAWNS
A garden that relies for effect on immaculately mown grass and a profusion of leafy, well-nourished plants is bound to be one of the first to suffer in prolonged periods of drought unless it can be given a supplementary supply of water. In areas where water is limited, a verdant lawn may have to be regarded as dispensable and replaced by gravel or other hard landscaping (see p.24).

OTHER ELEMENTAL EFFECTS

Next to rainfall, soil type has the greatest impact on the amount of water available to plants (*see p.20*). In areas of low rainfall, a garden with clay soil will usually support a different and much wider range of plants than a fast-draining sandy or gravelly soil. Soil depth also affects moisture retention; in chalky districts, there may be no more than 15cm of soil over the hard bedrock of limestone, severely limiting the amount of water that can be retained. Soils with a naturally high organic content, such as peaty soils, are less likely to be a problem, since peat retains water.

Although in many temperate regions wind is often associated with driving rain, its general effect is to desiccate foliage, especially if it is persistent. Not only does wind increase evaporation from the soil surface, it also dramatically increases the rate of transpiration, and this loss of water from leaves in turn places a greater demand for water on plant roots. Erecting windbreaks to counter prevailing winds or temporarily screening vulnerable new plantings will help (*see p.15*).

Sunshine and shade can both add their own problems. In bright sunlight, even when temperatures are on the cool side, the rate of photosynthesis (the process by which plants manufacture the food that fuels their growth) rises and more moisture is lost in the process. Glare caused by extensive paved areas and reflective, light-coloured walls exaggerates the effect of the sun. Trees help to create cool shade, but the soil underneath is likely to be dry because of the umbrella effect of their canopies and their thirsty, extensive root systems.

PAVING THE WAY
A strong design that depends on attractive structural features will demand far less water than one centred around thirsty grass and lush borders. Here, vertical screens and pergolas help to reduce the drying effects of wind; decking and gravel reduce evaporation from the ground, while drought-resistant plants are used to good effect in raised beds and timber planters.

TYPICAL TROUBLESPOTS

A BIRD'S EYE VIEW OF THE AVERAGE BACK GARDEN quickly reveals not only those features that demand, or waste, most water but also how opportunities for saving it can easily be missed. In this garden, large amounts of rainwater could be collected from the roofs of both house and shed and stored in a butt (*see p.45*). This, in turn, could be used to water containers, especially hanging baskets, and the vegetable plot, and to top up the pool if necessary.

WHERE THE WATER DISAPPEARS

A large percentage of water, of course, simply disappears into thin air – lost through evaporation. Surfaces, therefore, are one of the first areas to look at for ways of saving moisture. A good thick mulch (*see p.22*) laid between flowers, shrubs and vegetables while the soil is damp is invaluable for helping to conserve moisture in beds and borders. Mulches are just as useful, too, in containers where they can add to the decorative effect (*see p.39*).

Lawns are the thirstiest surfaces of all. Water is lost through transpiration from each and every blade of grass. Choose appropriate grasses (*see p.27*) and mow less frequently. Under trees it is often best to concede the battle and use drought-tolerant ground-cover plants (*see p.34*).

To stop rainwater disappearing from patios straight down the drain, give paved areas a gentle slope towards the lawn and, if appropriate, the pool. Permeable joints between slabs or bricks allow water to seep through and reach the roots of plants that extend beneath. Timber decking offers the same advantage (*see p.25*).

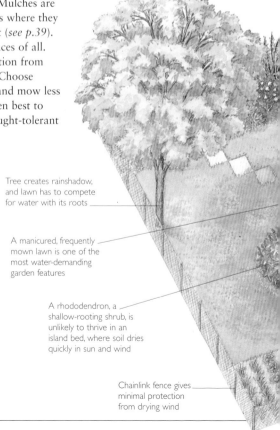

POTENTIAL PROBLEMS

• Wire fencing on boundary offers little protection from drying wind.
• Bare soil will lose moisture in sparsely planted beds.
• The lawn will be an early victim of drought and soon turn brown without supplementary watering.
• Hanging baskets and widely spaced containers need frequent watering.
• Fountain and pool will lose water, especially through evaporation.

Tree creates rainshadow, and lawn has to compete for water with its roots

A manicured, frequently mown lawn is one of the most water-demanding garden features

A rhododendron, a shallow-rooting shrub, is unlikely to thrive in an island bed, where soil dries quickly in sun and wind

Chainlink fence gives minimal protection from drying wind

APPROPRIATE PLANTING

Modifying the choice of plants helps to save water. Avoid notoriously thirsty trees like willow and poplar. Similarly, exchange runner beans for less water-demanding root vegetables and onions, and forego delphiniums in favour of drought-resistant flowers such as verbascums. Specimen plants in the centre of a lawn or in single containers on a patio lose more moisture than if they are part of a group planting. And all plants will benefit from some form of screen that reduces wind (*see p.15*).

SIMPLE REMEDIES

• Erect a type of fencing that will act as a windbreak, and install water butts.
• Mulch bare soil and start a compost heap. Use compost to improve soil's water retention.
• Do not cut grass too closely, and/or reduce lawn size or replace with a different surface.
• Incorporate water-retaining granules into the compost in pots and hanging baskets.
• Use a heavier, single spout fountain, or dispense with it and plant water lilies to cover pool surface and reduce evaporation.

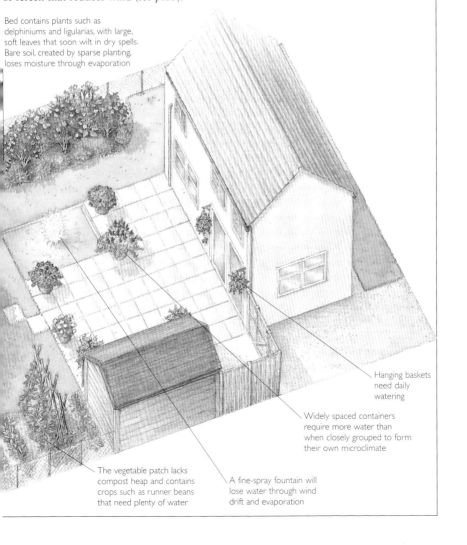

Bed contains plants such as delphiniums and ligularias, with large, soft leaves that soon wilt in dry spells. Bare soil, created by sparse planting, loses moisture through evaporation

Hanging baskets need daily watering

Widely spaced containers require more water than when closely grouped to form their own microclimate

The vegetable patch lacks compost heap and contains crops such as runner beans that need plenty of water

A fine-spray fountain will lose water through wind drift and evaporation

SOME BASIC SOLUTIONS

A GARDEN'S POTENTIAL FOR SAVING WATER can often be improved gradually, but some fundamental decisions may be required at the outset. Very windy sites, especially near the coast, are likely to need a hedge or shelter belt of trees and shrubs that takes time to establish (*see p.34*), although an easily erected artificial windbreak solves the problem in many gardens. Slopes may prove a more intractable problem, and on steep sites, terracing is the only answer.

COPING WITH SLOPES

On slopes, water runs off quickly and has much less opportunity to be absorbed by the soil. Where plants are not being planted in great number or in rows, form a shallow depression in the soil around each plant that can act like an individual reservoir (*see p.44*). This is particularly beneficial for shrubs and trees. On a steep slope, however, in order to create reasonably level planting areas, terracing may have to be carried out and, if necessary, retaining walls built. Where there is hardly any top soil and the bedrock is exposed, planting will need to be limited to rock plants, which are superbly adapted to grow in this

> ### Terracing gives the soil a chance to absorb as much rain as possible

type of terrain. Sometimes there is so little soil for the plants to gain a foothold that the only solution is to sow seed *in situ*, enabling the developing plant to squeeze its roots into any available fissures in the rock.

AT THE ROCK FACE
Sempervivums, adept at finding moisture in tiny cracks, are the perfect plant for clothing a rocky slope. After planting, some temporary netting will help to hold the plants in place.

WILLOW AND WOOD
Woven willow (left) *makes a good short-term windbreak, ideal for sheltering shrubs as they grow to form their own barrier. Spaced wooden laths* (below) *are effective and create dramatic shadows in winter.*

FILTERING THE WIND

Windbreaks of various forms should be considered carefully, particularly in temperate climates where drying wind is a factor all year. Even if windbreaks such as ranch-style fencing (*see below right*) are obtrusive at first, when used to support plants they add a new dimension to the garden planting.

Windbreaks are more effective when they filter wind rather than block it. A solid windbreak, such as a close-boarded fence, creates turbulence and eddying on the leeward side which can cause a surprising amount of damage. Semi-permeable artificial windbreaks, however, can be decorative in their own right. Trellis painted in a sympathetic colour makes a good garden feature at the same time as reducing wind. Any timber should be pressure treated and supported by strong pressure-treated posts. These can even be of concrete in extremely exposed positions.

The ideal permeability of a windbreak needs to be about 50 per cent. This can be achieved by making panels of vertical timber laths, approximately 2.5cm wide, with gaps in between of the same width. A windbreak's height is generally determined by the other features in a garden, but even a low trellis fence will reduce wind at ground level by some 7–10 times its own height. Added to this, there is the bonus that the windbreak can help to reduce the sun's glare, particularly if some small ornamental trees are planted alongside.

GOOD WINDBREAKS

• Panels made of vertical 2.5cm timber laths with 2.5cm gaps in between.
• Horizontal ranch-style (baffle) fencing, using 15cm wide timber with 15cm gaps.
• Trellis with a 15cm grid.
• Hazel or willow woven hurdles.
• Polypropylene windbreak mesh (available in various colours and thicknesses).
• Rolls of reed or split bamboo attached to upright frames.
• Evergreen hedges such as holly (*Ilex*).
• Deciduous hedges such as hawthorn (*Crataegus*) or sea buckthorn (*Hippophae*).

IMPROVING SOIL AND SURFACES

CONSERVING EVERY DROP OF WATER in the soil must be one of the main aims of the water-conscious gardener. This is best achieved by improving the soil's structure and covering its surface. Mulches may be restricted to beds and borders, but if you use a material such as gravel, which is decorative and easy to maintain, it can introduce a style to the garden that is so successful it does away entirely with the moisture-demanding, time-consuming lawn.

MAKING A GRAVEL GARDEN

Gravel blends beautifully with many drought-resistant plants and encourages a freer and more ecologically sensible gardening style. For maximum water conservation, lay the gravel over a sheet mulch (*see p.23*). But one of the charms of gravel is the way it allows plants to self-seed. For this to happen, simply spread it over the soil, 5–7cm deep, before or after planting. The plan overleaf uses plants that thrive in the conditions that gravel creates. The gravel soon becomes less dominant as the plants develop, their growth speeded by being given a cool root run. In winter, in some climates, it will also protect the lower leaves of grey-foliaged plants from excess winter wet.

1 Scoop the gravel from the planting area, if planting after the gravel is laid, and put to one side so that it does not fall into the planting hole.

2 Having made a good-sized planting hole, add some compost and lightly tease out the plant's roots. Plant it level with the surrounding gravel.

3 Fill the hole with soil, as necessary, and firm the plant in. Water it well. Brush the gravel back under the leaves and around the stem.

TEXTURAL COMBINATION *Giant crambe leaves contrast with santolina's tight button flowers.*

A GRAVEL GARDEN

Light-reflecting gravel sets off silvery Mediterranean plants to perfection. In this plan (shown in late summer) aromatic lavender and santolina will scent the air even in winter, while the euphorbia sprawls languidly near their feet. A prairie grass adds texture and sound when it rustles in the breeze. Although you will want some plants to self-seed, especially verbena, pull up unwanted tap-rooted eryngiums while small so as not to disturb the gravel.

The flowers of *Lavandula angustifolia* 'Twickel Purple' start to open after mid-summer and gradually fade as the season wears on.

Invaluable in a gravel garden, *Verbena bonariensis* has few leaves and thin, wiry stems. It is not fully hardy but self-seeds and can flower in its first year.

ERYNGIUM GIGANTEUM *Also called Miss Willmott's ghost, this plant has a luminous quality at dusk. Usually grown as a biennial, it self-seeds abundantly.*

Santolina pinnata 'Edward Bowles' forms a hummock of small lemon flowers from mid-summer on. Cut back in spring to retain its shape.

The spreading, fleshy, wine-red foliage of *Sedum* 'Bertram Anderson' contrasts with the harsh texture of the gravel.

In early summer, *Crambe maritima* has clouds of tiny white flowers that billow above the mound of glaucous leaves.

The leaves of *Calamintha nepeta* have a refreshing minty aroma, and the flowers attract bees.

PLANTING PLAN

1 2 × *Lavandula angustifolia* 'Twickel Purple' ♀, 60cm apart
2 3 × *Eryngium giganteum* ♀, informally arranged
3 1 × *Santolina pinnata* 'Edward Bowles' ♀
4 1 × *Sedum* 'Bertram Anderson' ♀
5 1 × *Crambe maritima*
6 1 × *Calamintha nepeta*
7 3 × *Verbena bonariensis*, 75cm apart
8 3 × *Schizachyrium scoparium*, 45cm apart
9 3 × *Perovskia* 'Hybrida', 60cm apart
10 1 × *Helianthemum apenninum*
11 1 × *Euphorbia myrsinites* ♀

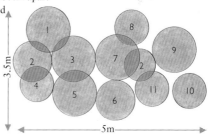

3.5m — 5m

MORE CHOICES

Ballota pseudodictamnus ♀
Cistus (especially small types)
Convolvulus cneorum ♀
Erysimum 'Bowles' Mauve' ♀
Festuca glauca
Limonium latifolium
Lotus hirsutus
Parahebe catarractae ♀
Stachys candida
Stipa tenuissima

Schizachyrium scoparium is a highly drought-resistant grass, native to the American prairies.

Perovskia 'Hybrida' has aromatic foliage. The blue flowers appear in late summer.

Self-sown eryngium seedlings will flower the following summer.

Helianthemum apenninum starts to open its papery white flowers in early summer. Regular deadheading helps to prolong its season.

EUPHORBIA MYRSINITES
With its trailing stems of blue-green leaves, this plant adds form and colour all through the year. The lime-green flowers in spring are a brilliant contrast.

LOOKING AFTER THE SOIL

HOW WELL YOU TREAT YOUR SOIL may be the deciding factor in whether plants survive in times of drought. Although soils vary in their ability to retain moisture, there are scarcely any that do not benefit from generous additions of organic matter. On fast-draining soil, it provides a plant's life-line.

ASSESSING YOUR SOIL TYPE

To understand how readily your soil might be affected by drought you need to establish its type. Sandy soil, with fast-draining large particles, dries out quickly. It feels gritty when rubbed, is often light in colour, and the grains will not stick together or form a ball. Thin, alkaline, chalky soil is almost as drought-prone. Often a pale greyish colour, especially when dry, it usually comprises a shallow band over limestone, and the surface can be quite sticky when wet. Clay soil, composed of fine, closely packed particles, retains water longer but becomes brick-like and unworkable when dry, with surface cracks appearing. When moist, it retains its shape if squeezed and feels slightly greasy. A cold soil, it is slow to warm up in spring.

PREVENTING COMPACTION

Compacted soil damages plant growth. It is unlikely to be a problem if you mulch and add organic matter on a regular basis.

SURFACE CAPPING
Walking on damp soil or overwatering can cause a hard surface crust or "cap" to form, preventing water (and oxygen) from reaching plant roots. Break up a cap if it occurs, but try never to tread on wet ground (if unavoidable, lay a board on the soil).

IMPERMEABLE LAYERS
Severe compaction creates a "hardpan" or almost impermeable layer under the surface; water cannot drain and roots get waterlogged. Break it up and add lots of organic matter.

WAYS OF IMPROVING MOISTURE RETENTION

Adding organic matter is the best way to improve water retention. Worked in among the soil particles, it acts like a sponge, holding on to moisture and making it available to plants when rainfall is short. On heavy clay, incorporate coarse grit as well as organic matter to open up the soil and prevent it cracking as it bakes hard in hot, dry conditions, damaging plant roots in the process. Dig in organic matter in winter to give it time to break down into a crumb structure by the growing season.

NATURE'S WAY
Organic matter is naturally taken down into the soil by worms and other beneficial micro-organisms, which break it up into an excellent soil conditioner. Avoid excessive tidiness in the garden and some of the work will be done for you.

THE BEST ADDITIVES

There are many forms of organic matter, some better suited to different types of soil. As a guideline, you will need to apply a layer at least 5–10cm deep to gain any real improvement. Add more to sandy soil and, for maximum effect, dig it in while it is damp. Recycling garden waste into compost makes one of the cheapest soil conditioners, and you can make leafmould by stacking fallen leaves in a chicken-wire enclosure to decompose. Most local authorities also now offer good, weed-free compost, recycled from municipal waste. Always compost fresh animal manure until it is well rotted.

ANIMAL MANURE
When well rotted it is a source of plentiful humus and some nutrients.

MUSHROOM COMPOST
Contains lime; do not use on very chalky soil or acid-loving plants.

GARDEN COMPOST
Wait until the texture turns crumbly. It is a source of humus and valuable nutrients.

LEAFMOULD
Good conditioner with low-level nutrients and humus. It suits acid-loving plants.

MAKING GARDEN COMPOST

Build up layers of soft, leafy material, such as lawn clippings, spent foliage and kitchen vegetable waste, and material such as tougher chopped prunings and straw. Bins can be of wood or plastic. If you have space, two bins are extremely useful so that you can add material to one while the other heap is rotting. Never apply compost to soil before the material is well-rotted – brown, crumbly and sweet-smelling. This usually takes about three months, but can take a year or more. Chop up coarse material (prunings and tough stalks) and mix with the soft material (see above) to speed decompostion. The process is also greatly speeded by adding fresh manure or a propietary activator at intervals. Cover with a piece of old carpet, to keep the heat in, then a lid. The heat created by the rotting process further fuels decomposition, but is rarely high enough to kill weed seeds, so put only leafy growth on the heap. Never add perennial weeds such as ground elder or bindweed, nor meat that can attract vermin.

BUILDING UP THE LAYERS
Add the material by spreading it in layers. Avoid thick wads of grass cuttings, which inhibit air flow. As the bin begins to fill, turn the contents to ensure an even breakdown.

THE IMPORTANCE OF MULCHING

MULCHES PLAY A VARIETY OF ROLES. They are essential in combating drought because they substantially reduce the amount of water lost from the soil by evaporation. Yet they also suppress weeds, can look decorative, and, as they get taken down, organic mulches used regularly will gradually improve the soil.

NATURAL MULCHES

Organic mulches such as chipped bark, rotted manure and cocoa shells rot with time and generally need replacing at least every year. The rate at which they disappear depends on source, age and particle size. Spread them at least 10cm deep onto moist ground. If possible, use materials with a large particle size; small, crumbly mulches, though good at conserving moisture, make a first-rate seed bed for weeds. Most organic mulches are ideally suited to shady areas. Gravel looks perfect in open, sunny gardens and is a natural companion for plants that thrive in these conditions.

CHIPPED BARK
Variable in quality and particle size. Ensure you buy bark, not chipped wood.

GRAVEL
Excellent, long-lasting inorganic mulch, available in a range of colours and sizes.

GRASS CUTTINGS
Compost with other material first. Never use grass freshly treated with herbicide.

COCOA SHELLS
A good light and porous mulch, with small amounts of nutrient.

SHEET MULCHES

Used on their own, unprepossessing sheet mulches are usually reserved for hidden vegetable patches or the allotment. But, combined with a layer of gravel or chipped bark (*see right*), they make a highly efficient mulch for the ornamental garden. There are two basic categories: non-porous types like black polythene, which provide warmth but, unless perforated, do not allow rainfall or air to penetrate, and porous fibre fleeces or woven polypropylene. These let in water and oxygen but at the same time suppress weeds. They are extremely durable but, once in place, they make it almost impossible to add organic matter or fertilizers to the soil.

MATERIAL OPTIONS

Bonded fibre fleece Purpose-made for mulching, long lasting and easy to lay. Allows water and air to pass through.
Woven polypropylene One of the most expensive, but available in heavier gauges than fibre fleece. Extremely long lasting and hard wearing; good under pathways.
Black polythene Cheap and useful on well-drained, very sandy or gravelly soils that need to retain maximum moisture.
Newspaper Can be quite effective if several layers are used and thoroughly soaked after laying. May only last one season.
Old carpet Cheap and effective for small areas but will eventually rot.

Laying Organic Mulches

Before laying a mulch, remove all perennial weeds, then cultivate the ground and, after about 10 days, hoe off any weeds that have germinated. Mulches are usually best laid in spring, while the soil is still damp from winter rain but has just started to warm up. Lightweight materials such as fine bark and cocoa shells are best applied on still days, or wind may blow away the particles. These mulches are also best watered immediately afterwards – water releases a gum in cocoa shells that binds the pieces together. If there are nearby paths or lawns, make a raised edge or kerb to stop the mulch from constantly spilling over.

THE BEAUTY OF BARK
Bark chippings, laid 10–15 cm deep, are ideal, as their coarse texture readily allows rain through without making them cake together.

Concealing Sheet Mulches

Sheet mulches must be laid before you begin planting but after the soil has been thoroughly prepared. After rolling out the sheeting, anchor it in position with plastic or wire pegs, or push the edges of the sheet into the ground with a spade. Put in the plants before spreading the decorative mulch. Make it at least 5 cm deep or the sheet may show through. If using gravel on large areas, you can add interest by varying the size of the gravel and including one or two larger rocks.

1 **To plant,** cut a crosswise slit in the sheet mulch and fold back the four corners. Make a comfortable-sized hole for the rootball. Put in the plant and water in well, then fold the sheet mulch back in place.

2 **Spread bark chippings** – or other ornamental mulch – over the sheet, brushing it under leaves and around plant stems with your hands. Ensure that the whole area is well and evenly covered.

BRIGHT ALTERNATIVES
Glass nuggets add lively colour in dreary seasons. Use in small quantity and choose plant partners carefully for a cohesive, not chaotic, effect.

GOOD GARDEN SURFACES

THE WATER-WISE GARDEN OFFERS enormous scope for using surfaces other than lawn. Paving, cobbles, gravel or decking can be combined in an imaginative way that conserves moisture at the same time. These materials keep evaporation to a minimum, need little maintenance and provide a cool root run for plants, and a subtle mix of textures and colours can look good even in a tiny garden.

ON THE GARDEN FLOOR

Gravel is one of the most versatile materials, ideal as a mulch and background for plants (*see p.18*) but also suitable for more open spaces such as sitting out areas. For these it needs to be laid on top of a firm subsoil base and layer of hardcore, to barely more than 2.5cm deep so that you can walk on it comfortably rather than wade through. Gravel often looks best – and is easiest laid – given a neighbouring hard surface or edge of cobbles, paving slabs, bricks or setts. In any case, do not lay it right up to the house entrance, since it is easily trodden indoors. When laying paving, bed bricks or slabs on sand, with just a few joints held in place with cement, so that rain drains through. Where a harsh climate necessitates sparse planting, tiles or glass blocks can introduce small areas of contrast. Whatever the material, choose colours and textures that blend with the house.

► DESIGN IN THE ROUND
Cobbles and gravel are easily worked into curving shapes and naturally complement one another as well as the plants.

◄ STONE AND GLASS
Glass blocks (far left) create clear pools of colour among rough-textured gravel but are not meant for areas of hard wear and tear. More durable concrete roundels (left) are reminiscent of a log path, but suit a dry garden, where shades of grey predominate instead of woodland green and brown.

DECKING BY DESIGN

Timber decking has a pleasantly flexible feel to walk on, and also allows rain to pass through to plant roots below. You can use a natural shade of preservative or introduce colour. Although decking can be set at the same level as the rest of the garden, it often looks better slightly raised, but supporting structures must be strong enough to take the weight. If cobbles or gravel are used as a surround, you can create the illusion of a jetty over a dry river bed.

MAKING IT SAFE

• Grooved timber helps to prevent slipping when the surface is wet. If algae forms, remove it with a fungicide and stiff broom. Special deck paints for yachts, which create a non-slip, gritty surface, can also be used.
• Use pressure-treated timber and supplement with a preservative. Do not allow joists to come into contact with soil.
• Put handrails at the side of any steps.

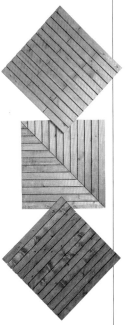

HARMONIOUS WOOD
Decking provides a sympathetic link between garden and home. Planks can create bold directional lines and angles, while ready-made panels come in a variety of patterns and are easily laid direct on level ground.

DECKING PANELS

PLANTING IN PAVING

Take the edge off hard-surface areas by filling the cracks between paving stones with creeping plants, such as thyme. Even in dry climates, there is usually sufficient moisture beneath the slabs for the efficient root systems of alpines to exploit this zone. The effect is enhanced when the occasional slab is left out and a taller or bushier plant allowed to fill up the space.

PAVING INTERPLANTED WITH THYME

EFFECTIVE GROUND COVER

ONE OF THE CHALLENGES in drought-resistant gardening is the creation of interesting plant combinations that require little maintenance yet provide efficient ground cover (if necessary, the plant roots helping to bind the soil). In many dry gardens, planting is sparse and concentrates on the bold use of key specimens. The art of good ground cover, however, lies in creating drifts of plants that are broken by an occasional change in height, shape and colour.

GETTING THE BEST RESULTS FROM PLANTS

The grey foliage of many drought-resistant plants acts as the perfect foil for darker leaf or flower colours, particularly purple. Aim for the bulk of the plants you choose to grow no more than 30–45cm tall. At this height they are unlikely to be damaged by wind. Choose plants with a spreading or hummocky habit, including a proportion of evergreens such as acaenas, alyssum, hebes and sage so that there is some cover and colour in winter, and arrange them in uneven-shaped drifts. Then, as they grow into each other, they will achieve a pleasing informality. At the same time, they will create a slightly undulating framework in which small gaps should be left for more upright or spiky plants – the accent plants, with strong foliage or striking colour.

PLANTING TIPS

• Incorporate as much organic matter, such as compost, as possible after clearing the ground of weeds, especially perennial weeds.

• Plant in the autumn to give plants a chance to establish before the dry season.

• Small, young plants establish more quickly in dry conditions than older, larger plants.

• Remove any leggy shoots to encourage the development of bushy, sturdy plants.

• Plants with roots to bind soil include *Atriplex*, *Elymus*, *Epimedium*, *Eragrostis*, *Helianthemum*, *Rosa rugosa* and thyme.

TAPESTRY OF PLANTS
Mounds of purple sage, lavender and thyme are punctuated by spires of verbascum, orange splashes of poppy and grassy leaves of carex.

GOOD GROUND COVER PLANTS

LOW-GROWING

Herbaceous and rock plants
Acaena, alyssum, *Antennaria*, *Arabis*, *Aubrieta*, *Cerastium*, *Epimedium*, *Saponaria*, *Sedum*, thyme

Shrubby plants
Cotoneaster horizontalis ✿, hebes such as *H. ochracea*, *Hedera* (ivy), *Helianthemum*, *Hypericum calycinum*

Ornamental grasses
Briza, *Festuca* (fescue), *Holcus*

MEDIUM HEIGHT

Herbaceous and rock plants
Artemisia, nepetas such as *N. × faassenii*, *Origanum*, *Pachysandra*, *Phlomis russeliana* ✿, salvias such as *S. × superba* ✿, *Senecio*, *Stachys*

Shrubby plants
Artemisia, *Cistus*, *Genista*, *Juniperus procumbens*, *Salvia officinalis* (sage), *Santolina*

Ornamental grasses
Elymus, *Helictotrichon*, *Koeleria*, *Stipa tenuissima*

TALL PLANTS

Herbaceous plants
Acanthus, achilleas such as *A.* 'Coronation Gold' ✿ and *A. filipendulina* 'Gold Plate' ✿, *Gaillardia*

Shrubby plants
Atriplex halimus, *Elaeagnus angustifolia*, lavender, *Perovskia*, rosemary, *Rosa rugosa*, *Ruta graveolens* (rue)

Ornamental grasses
Eragrostis, *Leymus*, *Melica*, *Pennisetum*

▲ *For details of hardiness, see Good Plants for Dry Places, starting on p.47.*

A NEW LOOK AT LAWNS

No widely available lawn grass mixtures can survive drought and yet remain green if closely mown. Although most lawns recover when rain returns, the main reason for rapid browning is mowing too closely – shaving off the plants' "food factory".

Keep the winter cut no lower than 25mm; in spring, reduce this to 13mm and give the lawn a feed. If the grass starts to brown, raise the height of cut to give only a light trim. When sowing lawn seed, try to use a mixture with a high proportion of fescues,

particularly hard fescue and slender red fescue. A good mix would be 60 per cent creeping red fescue, 15 per cent chewings fescue and 10 per cent each hard and slender red fescue and browntop bent. For a large area, that is cut no lower than 25mm, try 25 per cent each creeping red and hard fescue, 20 per cent smooth-stalked meadow grass and 10 per cent each slender red fescue and browntop bent. As an alternative, on sandy soil you could make a small lawn of prostrate thyme (*see T. serpyllum, p.67*).

THYME IS RIGHT
A thyme lawn doubles as a sundial. Prostrate thymes can withstand being walked on; it heightens the aroma from their creeping stems.

WHICH GRASS TO CHOOSE

Lawn grasses are split into cool season for temperate areas (preferring temperatures of 15°–25°C) and warm season for subtropical climates (preferring 26°–35°C).

COOL SEASON GRASSES

Slender red fescue; hard fescue; smooth-stalked meadow grass; and, for very dry conditions where a fine finish is not needed, crested wheatgrass (*Agropyron cristatum*) and Western wheatgrass (*Agropyron smithii*).

WARM SEASON GRASSES

Bermuda grass (*Cynodon dactylon*); *Zoysia*; St Augustine grass (*Stenotaphrum secundatum*)

CHOOSING THE RIGHT PLANTS

A GROWING ENVIRONMENTAL AWARENESS and appreciation of native landscapes, especially where they are under threat, is changing attitudes and encouraging gardeners to take a much more natural approach. Plants such as grasses are now highly valued for their easy charm, grace and ability to grow well without undue amounts of cosseting. Match plant with site and gardening becomes liberating not restricting, and the effects created are satisfying in every sense.

NATURAL SURVIVORS

For the best drought-resistant plants look at what grows well in the wild in comparable habitats and conditions – for example Mediterranean shrubs and bulbs, prairie plants and downland and seashore natives. Many have developed easily identifiable characteristics (*see p.32*), and often make excellent natural partners. Garden designers are drawing on this to produce inspirational combinations. Often called "new-style" perennial plantings (*see the plan overleaf*), these use plants appropriate to the conditions, which need no staking or winter protection. Once established such schemes need minimal care.

LIFE ON THE EDGE
Tucked between rocks on a sandy shore, thrift (Armeria) *manages to survive on the meagre amounts of moisture held in the crevices. Other plants such as* Tamarix, Crambe *and* Eryngium maritimum (*sea holly*), *which have little but sand to draw water from and are buffeted by salt-laden winds, are real drought survivors.*

◄ EASY-CARE PLANTS *Alliums, yellow asphodel and red euphorbia provide layers of colour and shape.*

NEW-STYLE PERENNIAL BORDER

Try to make the arrangement look as natural as possible – loose drifts help produce a continuous flow of interest. Alliums may slowly diminish, but do self-seed, and it is easy to add more bulbs if needed. The euphorbia's spent flower stems are best cut back, and deadheading the nepeta will encourage many more blooms, but leave the dried heads of achilleas, alliums and grasses for autumn and winter display. Here, the garden is seen in mid-summer.

ECHINOPS RITRO
The round flower globes, which appear in late summer, echo the shape of the alliums on a smaller scale. They are excellent for attracting bees and butterflies.

MORE CHOICES

FLOWERING PLANTS
Anthemis
Asclepias tuberosa
Dictamnus albus ♀
Gaillardia
Gypsophila repens ♀
Lychnis flos-jovis
Phlomis russeliana ♀
Salvia sclarea
Sedum
Verbascum

GOOD GRASSES
Cortaderia selloana
Elymus magellanicus
Eragrostis curvula
Melica altissima

The large, flat, dark red flowerheads of *Achillea millefolium* 'Sammetriese' start to open in mid-summer and are extremely long-lasting.

Nepeta x *faassenii* makes a greyish mound of aromatic foliage, covered from mid-summer with a succession of lavender-blue flowers.

Over the summer, the evergreen leaves of *Stipa arundinacea* gradually turn tawny, until in winter they are entirely russet-brown. A decorative plant all year.

PLANTING PLAN

3 × *Achillea millefolium* 'Sammetriese', 60cm apart
3 × *Echinops ritro*, 60cm apart
1 × *Stipa arundinacea*
3 × *Origanum laevigatum* ♀, 45cm apart
3 × *Echinacea purpurea* 'Robert Bloom', 60cm apart
3 × *Nepeta* × *faassenii*, 45cm apart
3 × *Helictotrichon sempervirens* ♀, 30cm apart
3 × *Euphorbia griffithii*, 60cm apart
4 × *Asphodeline lutea*, 30cm apart (small plants)
9 × *Allium cristophii* ♀, arranged informally

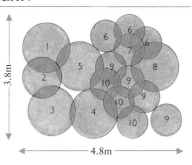

3.8m

4.8m

Helictotrichon sempervirens
has leaves in an eye-catching
shade of blue-grey; tall,
slender flowering stems are
produced in summer.

ECHINACEA PURPUREA
The red-purple flowers of
'Robert Bloom', with their
high-domed rusty cones,
provide useful colour in late
summer and autumn.

The flowerheads of
Euphorbia griffithii are an
unusual dusky orange-red
that harmonizes well with
other plants. It spreads by
underground runners.

Fat, bobbly seedcases
decorate the stems of
Asphodeline lutea after the
yellow star flowers have gone.

The spherical flowers of
Allium cristophii have impact
even in autumn, long after
they have faded and
turned brown.

ganum laevigatum bears
flowers on wiry stems
n late spring to autumn.

WHICH PLANTS CAN COPE?

PLANTS FROM NATURALLY DRY REGIONS have all had to evolve a strategy of one kind or another for surviving drought. Many have developed distinctive types of foliage (*see opposite*), often pleasing to the eye and easy to use in a decorative way. Roots, too, have to be capable of seeking out every scrap of water. Those plants that have adapted best to the dry life are known as xerophytes, and drought-resistant gardening is sometimes called xeriscaping.

FOLIAGE SURVIVAL TECHNIQUES

Generally the smaller a plant's leaf the more likely it is to be able to withstand drought, because little water is lost from the surface. Some grasses can roll their blades inwards, making them even narrower. In plants like broom, many or even all leaves have been lost, with stems taking over the job of photosynthesis and food manufacture.

Grey foliage is another characteristic. Light-coloured leaves reflect glare. Sometimes the greyness or silveriness is caused by a covering of hairs; these help to lower the temperature within the leaf, reducing moisture loss, and also limit the drying effects of wind. Glaucous foliage, such as that of eucalyptus, often gets its bluish tone from a waxy coating. Other leaves are protected by a leathery skin. And many plants combine techniques: for instance, the fleshy leaves of succulents, such as sedums, not only store water but also have a waxy surface.

UNDERGROUND SOLUTIONS

Many desert plants have developed long, deep tap roots to search for water at lower, cooler levels. Plants in shallow soil, especially over rock, produce fine, fibrous, very extensive root systems that can seek out water between each and every particle. Where drought is seasonal, bulbs and corms produce foliage and flowers in the cooler, moister times of year. In the dry season they go dormant, storing food underground to provide energy for new growth when the rains return.

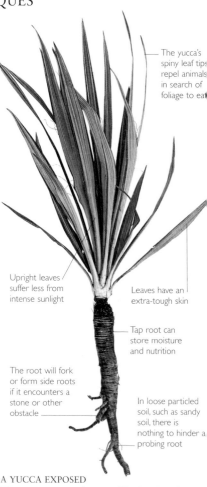

The yucca's spiny leaf tips repel animals in search of foliage to eat

Upright leaves suffer less from intense sunlight

Leaves have an extra-tough skin

Tap root can store moisture and nutrition

The root will fork or form side roots if it encounters a stone or other obstacle

In loose particled soil, such as sandy soil, there is nothing to hinder a probing root

A YUCCA EXPOSED
This young yucca has quickly developed a long tap root capable of plumbing the depths. Its leaves have a tough, waxy skin that retains moisture and can withstand extremes of heat.

FOLIAGE THAT CAN WITHSTAND DROUGHT

Eucalyptus perriniana
This eucalyptus is not wilting. Instead it has naturally drooping leaves that turn away from the sun in order to reduce the effect of its rays and mimimize water loss. (*See also p.48.*)

Sedum spathulifolium 'Purpureum'
A tiny succulent with several tactics. It can store water in its fleshy leaves; tightly clustered, they present only a small surface area to the sun and efficiently collect any moisture.

Portulaca oleracea
Known as purslane, this is cultivated in Mediterranean countries for its water-retentive, fleshy leaves, which give a refreshing bite to salads. It has some brightly coloured relatives for the flower border (*see p.69*).

Eryngium maritimum
The leaves of this seashore dweller have an extremely tough, leathery skin that inhibits moisture loss. It also has a long tap root that travels deep to find water in impoverished sandy soil (*see also p.60*).

Sempervivum ciliosum
A rock plant that combines strategies. The leaves curve inwards and are covered in sun-reflecting hairs (*cilium* means hair). It often grows among rocks to protect it from heat. (*See also p.67.*)

Senecio cineraria 'Cirrus'
A thick coating of fine hairs gives this plant its grey, felty appearance. Hairs often create an attractive silky or furry look, but also make the plants susceptible to a combination of winter cold and wet.

Festuca ovina
The blue-green grassy quills have a very small surface area, losing very little water. The roots form a fine, fibrous mass, maximizing the number of root hairs that can draw moisture from the soil.

Pinus thunbergii
Pines are among the most drought-resistant conifers. The needles have not only a small surface area but also a thick waxy skin that helps to seal in moisture and prevent them from drying out in wind. (*Other pines, see p.49.*)

PLANTS FOR DIFFICULT PLACES

As well as adapting to drought, plants in some sites have further problems to deal with. Dry shade under trees, especially evergreens, creates particular difficulties, often in gardens that are otherwise relatively damp. In coastal sites plants must withstand exposure to salt and wind, and in real desert, where annual rainfall is less than 25cm, survive exceptional extremes of heat and cold

DEALING WITH DRY SHADE

The combination of poor light, impoverished soil and tree or shrub roots makes dry shade one of the most limiting environments. Removing trees' lowest branches to let in light and improve air circulation encourages better plant growth. Spring-flowering bulbs that go dormant during the driest period are usually a good choice for such a site.

LAMIUM MACULATUM

PLANTS FOR DRY SHADE

HERBACEOUS PERENNIALS
Bergenia, Brunnera, Cortaderia selloana (pampas grass), *Epimedium**, *Lamium**, *Pachysandra**, *Pulmonaria, Waldsteinia**

BULBS AND CORMS
Allium, Anemone blanda ♥, *Cyclamen coum* ♥ and *C. hederifolium* ♥, *Eranthis hyemalis* ♥

SHRUBS
Cotoneaster horizontalis ♥, *Euonymus, Hedera* (ivy)*, *Ilex* (holly), *Prunus laurocerasus* (cherry laurel) ♥, *Ribes* (flowering currant), *Ruscus* (butcher's broom)*, *Santolina, Symphoricarpos, Vinca minor* (periwinkle)*
*makes good ground cover
Dry shade plants are often unsuitable for sun.

BY THE SEA

Fast-draining sandy soil and strong drying winds are a real problem in coastal gardens. Hedges of salt-resistant trees and shrubs help to shield other plants. When choosing garden plants, look at what grows wild along the shore for a useful guide. Seaside natives withstand drought because they have had to develop a physiology that keeps out sea water. A plant name ending in *maritim* or *maritimum* is also a good indication.

GOOD PLANTS FOR COASTAL SITES

FLOWERING PERENNIALS
Armeria (thrift), *Cortaderia* (pampas grass), *Crambe, Eryngium, Kniphofia* (red hot poker), *Limonium* (sea lavender), *Oenothera*

FLOWERING SHRUBS
Cistus, Cytisus (broom), fuchsias, *Genista*, hebes, hydrangeas, *Olearia, Romneya coulteri* ♥, × *Halimiocistus*, rosemary, *Spartium*

SEASIDE HEDGING PLANTS
Crataegus (hawthorn), *Elaeagnus* (evergreen types), *Escallonia, Fuchsia* 'Riccartoni' ♥, *Griselinia littoralis* ♥, *Hippophae rhamnoides* ♥, *Ilex* (holly), Leyland cypress, *Olearia* × *haastii, Pyracantha, Rosa rugosa, Tamarix* (tamarisk), *Ulex* (gorse – makes a low hedge, if clipped)

INTO THE DESERT

The arid areas of the world produce the most dramatic, sometimes bizarre, of all plant adaptations for storing water or minimizing its loss. A desert dweller like the saguaro cactus can store up to eight tons of water in the reservoirs of its thorny towering columns. Many xerophytes have vicious spines – added protection against grazing animals. Plants such as *Parkinsonia* (*see p.49*) may resort to shedding leaves (though this can happen, too, in temperate climates). The north American ocotillo (*Fouquieria*) keeps its leaves for a few weeks only before returning to being a bundle of thorny stems. The wide spacing between desert plants is no accident since it prevents competition for water reserves. In the south Arabian desert, four-fifths of some plants lies beneath the ground, the roots tapping into moisture stored up to 30m down.

▲ SPIKES AND BOULDERS
The juxtaposition of shapes is one of the most important elements in arid landscape gardens. Carefully placed rocks surround this agave.

▼ THE DESERT GARDEN
Flowering may be brief but for architectural form few plants compare with desert natives (see plan on next page for a desert garden).

A DESERT GARDEN

If you can give them the right conditions, desert plants present the opportunity to combine some of the most arresting, architectural shapes in the plant kingdom. As in a natural arid landscape, rocks and boulders are an important element. Most of the plants are tender (*see pp.47–69*), but a similar style could be created in a cooler climate if the plants were plunged in pots for summer, then lifted and protected under glass in winter.

PLANTING PLAN

1 3 × *Yucca whipplei*, 75cm apart
2 3 × *Lampranthus haworthii*, 30cm apart
3 1 × *Lantana montevidensis*
4 1 × *Aloe ferox*
5 1 × *Opuntia robusta*
6 1 × *Echinocactus grusonii*
7 2 × *Atriplex halimus*, 60cm apart

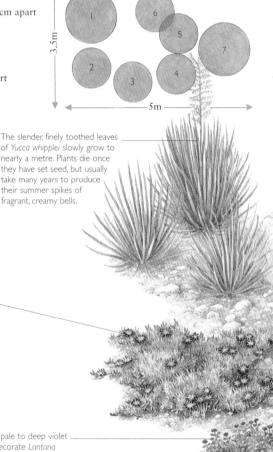

3.5m

5m

The slender, finely toothed leaves of *Yucca whipplei* slowly grow to nearly a metre. Plants die once they have set seed, but usually take many years to produce their summer spikes of fragrant, creamy bells.

LAMPRANTHUS HAWORTHII
A succulent plant that trails over the ground, its grey-green, fleshy foliage studded with eye-catching, purplish-pink daisy flowers.

Heads of pale to deep violet flowers decorate *Lantana montevidensis* during summer, standing up well above the leaves. The slender stems of this spreading shrub will form a dense mat.

CONSERVATION ALERT

• Some plants, including many cacti and other succulents, are under threat of extinction in the wild because of habitat destruction and over-collection. Gardeners can help to preserve endangered species by checking sources when they buy.

• Slow-growing plants such as cacti, which take several years to reach landscape size, are economically unfeasible to produce. When large specimens of such plants are offered for sale, especially at low prices, it means they have almost certainly been collected from the wild. Don't buy them unless the supplier can prove that the plants have been nursery-raised (usually from seed) or legally rescued from a construction site or similar area where they would otherwise be destroyed.

ECHINOCACTUS GRUSONII Often called the golden barrel cactus, the spiny, ribbed spheres very slowly become more elongated in shape. They are decorated with rings of yellow flowers in summer.

Opuntia robusta (prickly pear) is a slow-growing cactus. Its yellow flowers, produced in late spring and summer at the tips of the flattened segments, are followed by deep red fruit.

Like many desert natives, spiny *Aloe ferox* has a rosette of leaves that channel all the moisture it receives towards the roots.

The silvery, diamond-shaped leaves of *Atriplex halimus* (tree purslane) make it an attractive foliage shrub that will also suit more conventional situations.

MAKING THE BEST USE OF WATER

PLANTS IN CONTAINERS inevitably need regular watering, but by choosing the most suitable plants (*see plan overleaf*), containers and compost (*see p.42*) you can reduce the amount required. Similarly, some methods of watering the garden are more efficient than others (*see p.44*). Plants may have to rely on stored rainwater in spells of drought and during hose-pipe bans. Then, it may also be necessary to recycle waste domestic water, or "grey" water (*see p.45*).

GARDENING IN CONTAINERS

Many drought-resistant plants are suited to life in a container since they can tolerate a degree of neglect. Some of the most exciting are also the least hardy, but containers give cool-climate gardeners the chance to grow them, provided plants can be protected in a greenhouse or conservatory in winter. Water-retaining crystals and non-porous containers (*see p.42*) reduce the amount of watering needed, as do all kinds of mulches (*see p.22 and below*).

FALLEN BOUNTY
Pine cones, gathered from the woodland floor, make an unusual but natural-looking mulch. Match cone size with the size of the pot.

INSPIRED BY THE SEA
Combined with the right plant, a mulch of shells, pebbles and sparkling glass nuggets will help to conserve moisture in the compost.

TWO OF A KIND *Cordylines and agaves are good partners for each other and for flowering plants.*

A CONTAINER GARDEN

A group of containers of varying shapes and heights perfectly displays foliage textures and forms. Some of these plants (shown here in late summer) will need winter protection in cool climates. Brilliant gazania flowers circle the upright purple spears of a cordyline, and a sedum and *Convolvulus cneorum* trail naturally over the sides of their pots, as do the mats of foliage formed by the neat rosettes of two sempervivums.

PLANTING PLAN

1 2 × *Sedum* 'Ruby Glow' ♀, planted close to pot edges
2 1 × *Hebe* 'Red Edge' ♀
3 3 × *Sempervivum montanum*
4 2 × *Agapanthus* 'Blue Moon'
5 1 × *Cordyline* 'Purple Tower'
6 3 × *Gazania* Harlequin Hybrids
7 2 × *Convolvulus cneorum* ♀
8 3 × *Sempervivum giuseppii*

1.8m

3m

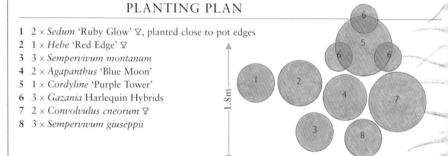

One of the lower-growing agapanthus, 'Blue Moon' is ideal for a pot. It can be moved into a shed for winter protection in cool regions once the foliage has died down.

SEDUM 'RUBY GLOW'
This small sedum has a naturally sprawling habit and suits life in a container, where its flowers spill over the sides.

With its purple-tinged young leaves, *Hebe* 'Red Edge' is a handsome shrub even without its lilac flowers. These open in mid-summer and gradually fade to white. It will need winter protection in frost-prone areas.

Rosettes of *Sempervivum montanum* multiply to cover a shallow pot. Occasional thick red flower spikes shoot up from their centres in summer.

Cordyline australis 'Purple Tower' adds architectural focus. In cold climates it needs the protection of a greenhouse or conservatory in winter.

MORE CHOICES

GOOD FOLIAGE

Aloe ferox, Cotoneaster horizontalis ♥, *Festuca glauca, Helichrysum petiolare* ♥, *Lotus hirsutus, Opuntia robusta, Pennisetum alopecuroides, Phormium tenax* ♥, *Yucca filamentosa* ♥

ATTRACTIVE FLOWERS

Brachyscome iberidifolia, Dimorphotheca pluvialis, × *Halimiocistus wintonensis* 'Merrist Wood Cream' ♥, *Halimium* 'Susan' ♥, *Lampranthus haworthii, Nerium oleander, Phlomis fruticosa* ♥, *Portulaca grandiflora, Saponaria*

AROMATIC PLANTS

Lavender, origanums, rosemary, *Salvia sclarea*, thymes

Eye-catching *Gazania* flowers are produced all summer but, like several African daisies, only open in sun. Best grown as an annual in temperate climates, it also makes a good seaside plant. Available in a wide range of colours.

CONVOLVULUS CNEORUM
A shrub that revels in sun. Its main flowering period is from late spring to mid-summer, but it often produces an occasional fluted trumpet later in the season.

The tight, rounded rosettes of vigorous *Sempervivum giuseppii* are flushed dark red at the tips.

WATERING CONTAINERS

MANY OF THE SAME rules apply to watering containers as to watering the garden (*see p.44*). Water well at planting, settling compost by using a fine rose on the can. Then, water thoroughly but only when plants need it, at the coolest times of day (water compost, not foliage). Use mulches, and even though you are trying to conserve moisture, ensure containers have good drainage holes.

CHOOSING THE BEST CONTAINERS

Terracotta sets off plants to advantage but, since it is porous, compost dries out fast. Paint or seal pots inside with varnish, or line with plastic punctured with drainage holes. Plastic and fibre-glass pots retain moisture well but can get hot in the sun. Standing them inside another container, such as a wooden planter, helps to insulate plant roots from the heat. Reconstituted stone keeps roots cool in hot spots. The larger the volume of compost, the slower it dries out, so choose big containers when possible. If you group pots, they protect one another from heat and drying wind.

PERFECT FINISH
Glazed ceramic pots are a good choice and come in lovely colours. They need drainage holes and must be frostproof in cool climates.

COMPOSTS

Multi-purpose composts, based on peat, dry out fast and are difficult to re-moisten. If choosing a peat substitute-based compost, opt for a moisture-retentive material such as coir. As well as adding crystals (*below*), incorporate some fine grit to aid rapid saturation when watering. Loam-based composts (John Innes composts) are slower to dry out. Leave plenty of space for watering between compost and pot rim and catch excess in a tray (replacing it with raised blocks in winter for good drainage).

ADDING WATER-RETAINING CRYSTALS
Specially formulated crystals swell with water and act as micro-reservoirs, slowly releasing moisture into the compost. Add before planting and use only the amount recommended.

WATERING IN WINTER

Container plants, especially evergreens, may sometimes need watering in winter. Wind has a very drying effect. Keep a check and water in mild spells only, never if frost is forecast.

IRRIGATION SYSTEMS FOR CONTAINERS

Automatic and semi-automatic irrigation systems can deliver water efficiently to a group of containers – straight to the compost and the plants' roots – as well as saving time. It is usually better to put together a system to suit your own needs rather than buy a standard multi-purpose kit. A "dripper", or "drip-head", is used to supply each container. Each dripper is fitted to a small-diameter, flexible pipe (the microtube), which in turn fits into a larger supply pipe connected to an outside tap. Simple push-on plastic fittings are used to make the connections, and adaptors and elbow connectors aid installation and help in running the supply pipe around corners.

Fixed output dripper

Microtube

Adjustable dripper

Supply pipe

WHICH TYPE OF DRIPPER?
Fixed output drippers trickle out the water at a pre-determined rate. Where plants need differing amounts of water, choose adjustable drippers so that you can vary the flow.

INSTALLING THE SYSTEM

• Cut the supply pipe to length once you have arranged your containers, and connect the microtubes to it at suitable intervals. Most systems include a punch for puncturing the correct size hole in the supply pipe. Put a stop-plug in the end.

• Large pots (especially if they contain a collection of plants) often need two or more drippers to ensure even watering. To decide the number of drippers, measure how much water each container requires using a watering can (*see below left*) and the amount each dripper delivers (sometimes given in the manufacturer's instructions). If necessary, add in extra drippers when you have seen how the system is operating.

• Check regularly that the system is delivering water efficiently, and clean the nozzles when necessary as they are liable to become clogged with compost.

FULLY AUTOMATIC SYSTEMS

Computerized automatic systems are expensive but make it possible to turn on the water at any time (including when you're asleep or away from home). The most sophisticated employ sensors to ascertain soil moisture levels, but for a container garden it is better to work out the individual requirements of each container according to its size and use an automatic timer in conjunction with adjustable flow drippers. To set the timer, measure the amount of water needed daily by each container using a watering can, then measure the amount discharged by the dripper over a given time.

COMPUTERIZED TIMER   SIMPLE TIMER

TIMING THE FLOW
A computerized timer (above left) releases water on a programmed setting; a simple timer (right) turns water off after a pre-set period.

WATERING FOR MAXIMUM EFFICIENCY

For watering to be effective it needs to be given in the right quantity and at the right time of day, season and the plant's own life cycle. Sprinklers are seldom the answer because they use a lot of water in an inefficient way, resulting in run-off and capping of the soil surface. Water thoroughly when needed; do not give small, frequent amounts, which encourages surface rooting.

GIVING PLANTS A GOOD START

The most important time to water is at planting and immediately after, especially when putting container-grown plants into open ground. Before planting, thoroughly wet the rootball by standing the plant, in its pot, in a bucket of water until the air bubbles stop rising. After planting, water in well, and in dry conditions continue to water for the next few weeks, especially in spring and summer. The plant should then be making new growth, including new roots. Ease off watering or it will stifle root development and make the plant dependent on watering. Do not plant in hot, dry spells.

RESERVOIR EFFECT
A "saucer-like" depression around the plant, with the soil mounded into "mini" walls, acts like a reservoir and prevents water from running away, especially when planting on a slope.

RIGHT DIRECTION
By inserting a pipe at the side of the planting hole, water can be channelled towards deep soil, encouraging the roots to extend in the right direction in their search for moisture.

EFFECTIVE SYSTEMS

For borders and vegetable and fruit plots perforated hoses (known as "leaky" pipes or "soaker" or "seep" hoses) are effective at delivering water accurately, straight to the soil and with a minimum of evaporation. Made of plastic or rubber that is designed to emit water either along the whole length or from tiny holes at short intervals, the hose is threaded among plants close to the ground surface. For preference, run the hose through a layer of mulch to avoid holes becoming clogged with soil.

▲ DRIP-FEED THE SOIL
Water oozes into the soil, above plant roots, along the entire length of a "seep" hose.

▶ MAKE USE OF A MULCH
Laying a perforated hose in a gravel mulch stops holes from becoming blocked.

WHEN TO WATER

In summer, water in early morning or evening, when evaporation rates are low and the water has a chance to soak into the soil. Some plants need more water at certain stages of their life cycle. Young ornamental trees are more likely to be damaged by a shortage of water in the first half of summer than in the second. Fruit trees should not go short while the fruits are swelling. This applies, too, to cane and bush fruit, and to vegetables such as potatoes, courgettes and peas while their fruits or tubers are developing. Avoid waiting to water until plants actually wilt, especially vegetables, or irreversible damage may occur. Do not let tomato plants go short of water while in flower. It can cause blossom-end rot, which damages the fruit.

RECYCLING WATER

Rainwater is easily recycled by connecting a water tank or butt to a down pipe. Make use of garage and shed roofs as well as the house roof. If the water is covered and free from algae it will last for six months. The average rainwater butt stores 120–200 litres. Because of the weight of water, it must stand on a firm, secure base. Volume can be increased by linking butts with a short pipe inserted into the overflow sockets, or you can buy larger tanks. (Thoroughly clean any tank that has previously been used to store another fluid.) A small submersible pump connected to a hose boosts delivery to distant parts of the garden.

GREY WATER

This is the term used for domestic water that has been used for washing up or bathing. Given that the average bath uses 120 litres, it can be well worth recycling for the garden.
• Store in a dedicated container and don't mix with other water.
• Use as soon as possible after it has cooled.
• Bath-water is more suitable than used water from the kitchen.
• Do not use water containing bleach or strong detergent.
• Do not apply directly onto plant leaves, but water the adjacent soil or container compost.
• Try to rotate the types of water given to any one area, and avoid the persistent use of grey water on one patch of ground.
• Do not use with a "dripper" system to avoid nozzles clogging.

SAVING RAINWATER

To estimate the water that could be saved from a roof, multiply the surface area, measured in cubic metres, by the annual rainfall (measured in metres) by 1,000. With an annual rainfall of 60cm, a garden shed roof measuring 2.4×3.6m would collect just over 5,000 litres; a house roof measuring 16.4×7.3m would collect nearly 72,000 litres.

WATER BUTT

Cover for safety and to keep water clean

Down pipe from gutter

Overflow pipe

Tap at level that allows watering can to fit beneath

Secure stand built of bricks or insulation blocks. Purpose-made plastic stands are also available

GOOD PLANTS FOR DRY PLACES

THE FOLLOWING PLANTS can all cope with a lack of water. Most are adapted to life in the sun, but a few like shade. Tender plants from hot, dry climates can sometimes survive cold but succumb when it is combined with the damp of a northerly winter. Symbols indicate each plant's preferred growing conditions.

◩ *Prefers full sun* ◪ *Prefers partial shade* ▩ *Tolerates full shade* ✳✳✳ *Fully hardy (down to -15°C)* ✳✳ *Frost hardy (down to -5°C)* ✳ *Half hardy (down to 0°C)* **Min. °C** *Indicates minimum temperature for tender plants* ♀ *RHS Award of Garden Merit*

GARDEN TREES

TREES PROVIDE SHADE and some, especially conifers, help to lessen the drying effects of wind. Many of these trees will, if planted in groups with shrubby infill, make a shelter belt. With shelter, the garden becomes a more pleasant place to sit and offers conditions where more vulnerable plants can thrive; this is particularly true in seaside areas where salt winds severely limit plant choice.

Acacia dealbata ♀ (Mimosa) Evergreen tree with deliciously fragrant heads of small, fluffy, round yellow flowers, borne at the branch tips from winter to early summer. The silvery, fern-like leaves resist moisture loss. Like many woody plants in the pea family, it resents hard pruning. Can be grown in favoured climates, given the shelter of a wall. Height varies, depending on growing conditions, from 5m to 30m. ◩ ✳

Arbutus (Strawberry tree) Small range of spreading, often shrub-like evergreen trees with attractive reddish

Fruits

Flowers

ARBUTUS UNEDO

peeling bark. *A. unedo* ♀ produces small, white or pink, pitcher-shaped flowers in autumn, its strawberry-like

fruits not ripening until the following year. Slightly less hardy, *A. andrachne* flowers in spring, with fruits ripening in the autumn. Both reach up to 8m and can be grown on alkaline soil, unlike much taller *A. menziesii* ♀, which needs acid conditions. ◩ ✳✳✳

Cedrus (Cedar) Handsome conifers for gardens with plenty of space. Wind and drought-resistant, and generally conical in shape, they can reach 40m high with a 10m spread. Cedars of Lebanon make imposing specimens on

CORDYLINE AUSTRALIS 'VARIEGATA'

sweeping lawns or terraces. *C. libani* 'Sargentii' is slower growing, with weeping branches, and more suitable for a small garden.
❂ ✱✱✱

Cordyline australis ❦
(New Zealand cabbage palm) An architectural, palm-like tree, up to 10m tall. Long, arching green leaves may be variegated. In cold areas, grow in a container (which will limit its height to 3m) and overwinter in a greenhouse or conservatory. The leaves of outdoor plants can be protected by wrapping as for phormiums (*see p.62*).
❂ ✱
C. australis, see p.38
'Purple Tower', *see p.41*

Crataegus (Hawthorn)
Thorny, mainly deciduous trees and shrubs, well able to survive in inhospitable sites, particularly as hedging. *C. laciniata* is ideal for an exposed dry garden, especially by the sea. Leaves are glossy and deeply lobed; clusters of white flowers in late spring

and early summer develop into showy, red fruits in the autumn. Makes a good specimen up to 8m tall.
❂ ✱✱✱

Cupressus (Cypress)
Evergreen conifers that form a dense, bushy screen – ideal for windbreaks or tall hedges in mild seaside areas. *C. macrocarpa* can reach 30m with a spread of 4m, increasing with age, or can be grown as a hedge. Its lemon-scented leaves are a sombre dark green. *C. sempervirens* makes the thin, 20m-high spires that punctuate Mediterranean hillsides. 'Stricta' ❦, growing to 3m, is almost pencil thin and very effective in formal gardens.
❂ ✱✱/✱✱✱

Eucalyptus (Gum tree)
Wide range of species, mostly drought-tolerant and with aromatic, grey-green leaves, though juvenile and adult leaves may differ from each other in shape and colour. Many are not hardy. With age, *E. gunnii* ❦ develops a spreading shape and colourful

CRATAEGUS LACINIATA

peeling bark. Can be regularly hard-pruned to restrict its size and retain its rounded, silver-blue juvenile leaves. *E. parvifolia* ❦ (to 15m) can be grown on shallow chalk soil but is not completely hardy; *E. pauciflora* subsp. *niphophila* ❦ makes a hardy small tree (to 6m), its stems covered in white, waxy bloom. *E. perriniana* also remains fairly small (4–10m), with long, drooping adult leaves, but it is not completely hardy.
❂ ✱/✱✱✱
E. perriniana, see p.33

Ilex aquifolium ❦
(Common holly) Spiny-leaved, dense, evergreen small tree or shrub, good in hedging or when grown as a specimen. Separate male and female plants, with only females producing the bright red, orange or sometimes yellow berries. Plenty of yellow- or cream-variegated types to choose from; the degree of prickliness of the leaves can vary greatly. Height also varies depending on growing conditions.
❂ ✱✱✱

EUCALYPTUS GUNNII

Koelreuteria paniculata ♀
(Golden rain tree)
Fine specimen tree, reaching
10m or more, with deeply
divided leaves that are
pinkish-red as they emerge,
become green and turn butter
yellow in autumn. Large
spikes of small yellow flowers
in summer are followed by
rosy, bladder-like fruits in hot
summers. Tolerates drought
and wind, but not if salt-
laden. Pruning is best avoided.
❐ ✳✳✳

Ligustrum lucidum ♀
(Chinese privet)
An evergreen small tree or
large shrub, occasionally
reaching 10m, with glossy
leaves and spikes of white
flowers in late summer,
followed by black fruits.
Makes a good specimen plant.
❐ ✳✳✳

Parkinsonia aculeata
(Jerusalem thorn)
A small, tender, deciduous
tree or large shrub from
southern US and Mexico.
Spiny shoots bear delicate,
divided leaves that fold up at
night (they will also drop

PINUS PARVIFLORA

KOELREUTERIA PANICULATA

during very hot, dry spells),
and bright yellow flowers in
spring. Can reach 10m tall.
❐ Min. 5°C

Pinus (Pine)
Wide range of evergreen
conifers. Several pines do well
in dry, windy and seaside
sites. Many make large trees
(for instance, *P. parviflora* ♀
grows 10–20m) but look for
dwarfer forms more suitable
for small gardens. *P. mugo*
(dwarf mountain pine)
reaches only 3.5m, and the
shore pine, *P. contorta*, stays
shrubby on poor sandy soil.
P. radiata ♀ and *P. sylvestris* ♀
make good shelter-belt trees.
P. thunbergii makes a
rounded tree up to 25m tall
that will tolerate salt spray.
❐ ✳✳✳
P. thunbergii, see p.33

Quercus (Oak)
Huge range of mostly very
large trees requiring a variety
of conditions. *Q. ilex* ♀ (holm
oak) is an evergreen with
tough leaves, glossy above
and grey-hairy underneath,
that are well equipped to
withstand drying winds. It
can reach 25m in height with

a 20m spread. *Q. rubra* ♀
(red oak) reaches similar
proportions. Fast-growing
and deciduous, its leaves turn
yellow to reddish-brown in
autumn. A good specimen
tree needing lime-free soil.
❐ ✳✳✳

Rhus typhina ♀ (Stag's horn
sumach)
An upright, deciduous, small
tree or large shrub, up to 5m
tall, that gets its name from
its velvety shoots. Divided
leaves turn flame-coloured in
autumn. Yellow-green flowers
in summer are followed by
cone-like clusters of crimson
fruits on female plants. Spreads
by suckers; can be invasive.
❐ ✳✳✳

Robinia pseudoacacia ♀
Fast-growing suckering tree,
up to 25m, with deeply
divided leaves. Strings of
fragrant, usually white pea
flowers are followed by
brown seed pods. 'Frisia' ♀
has golden foliage turning
orange in autumn, but no
flowers. 'Idaho' grows to 12m
tall and has dark pink flowers.
❐ ✳✳✳

RHUS TYPHINA

Shrubs and Climbers

Shrubs provide a garden's backbone, extending the season of interest. Some are suitable for hedges and screens, sheltering vulnerable plants from drying wind. Like trees, many drought-tolerant shrubs come from the Mediterranean and similar climates, and may suffer in cold, damp winters. It is easier to establish young plants in a dry garden and these will make better plants in the long term.

Artemisia

Shrubby artemisias, excellent for hot, dry sites, include *A. abrotanum* ♀ (lad's love), an erect plant (to 1m) with semi-evergreen, pungent, green-grey leaves, and even more feathery, silvery, 60cm-high *A.* 'Powis Castle' ♀. Both need cutting back in spring to stop them getting untidy.
❏ ✻✻/✻✻✻

Atriplex halimus (Tree purslane)

Good seaside shrub that can withstand drying, salt-laden winds. Semi-evergreen, with leathery, silver-grey leaves. Tiny greenish-white flowers appear in late summer and autumn. Grows to 2m high with a 2.5m spread.
❏ ✻✻
A. halimus, see p.37

Ballota pseudodictamnus ♀

Shrubby, semi-evergreen plant forming a low, 45cm mound. Grey-green leaves on woolly stems resist drought by curling inwards. Whorls of pinkish-white flowers appear late spring and early summer.
❏ ✻✻✻

Bougainvillea glabra ♀

A brilliantly coloured evergreen climber for a dry site in a warm climate, where it can reach 8m, its white or magenta flowers covering walls, arbours and pergolas. In cool climates it needs a conservatory.
❏ ✻

Caragana arborescens (Pea tree)

An upright, thorny deciduous shrub, up to 6m tall, with divided, light green leaves. Pale yellow pea flowers appear in late spring. 'Nana' has a dwarf, congested habit with twisted shoots, growing to only 1.5m high.
❏ ✻✻✻

Cistus (Sun rose)

Evergreen shrubs with a succession of short-lived flowers all summer. Leaves, often aromatic, are frequently grey-green. *C. ladanifer* ♀, one of the most drought resistant, has sticky leaves and crimson-blotched white flowers. It grows to 2m tall. *C.* × *purpureus* ♀, with dark pink flowers with a maroon blotch, makes a bush 1m tall and wide. Trim back after flowering to encourage its rounded shape. *C. salviifolius* 'Prostratus' is low and spreading, no higher than 25cm, with yellow-centred white flowers. Cistuses can be rather short-lived, especially on chalky soil.
❏ ✻✻

Convolvulus cneorum ♀

A compact, bushy shrub, about 60cm high, with silky, silvery leaves. White, funnel-shaped flowers emerge from clusters of pink buds from late spring to early summer. Enjoys baking sun and well-drained soils, and will not survive damp combined with

ARTEMISIA 'POWIS CASTLE'

CARAGANA ARBORESCENS 'NANA'

COTONEASTER HORIZONTALIS

cold. Can be grown in a container, allowing it to be given shelter from winter wet.
☐ ✳✳✳
C. cneorum, see p.41

Cotoneaster
Range of evergreen and semi-evergreen shrubs including tall types for walls or screens and ground-cover plants. Most have inconspicuous flowers followed by masses of berries. Deciduous *C. horizontalis* ♥ is about 1m high with a wider spread and branches that form a herringbone pattern. In autumn, the plant is covered with red berries that attract birds. *C. conspicuus* forms a dense mound, up to 1.5m tall, of evergreen leaves; its red berries last well into winter. *C. lacteus* ♥ is good for a tall evergreen hedge (trim lightly in late winter); berries follow the flowers.
☐ ✳✳✳

Cytisus (Broom)
Deciduous shrubs, varying in size, with prolific clusters of pea flowers in mid- and late spring (see also *Genista*). Flowers are usually cream to bright yellow, occasionally

pink or red. *C.* × *praecox*, about 1.2m tall, has arching stems; *C.* × *kewensis* ♥ is about 30cm high and spreading – excellent trailing over banks and retaining walls. All usually succeed on poor soils and must not be pruned hard when mature.
☐ ✳✳✳

Echinocactus grusonii
(Golden barrel cactus)
A round cactus, very slowly achieving greater height (about 60cm) than its girth. Angular ribs are edged with yellow spines, and it produces bright yellow flowers. It sometimes survives in frost-free gardens. To add drama to cooler gardens, grow it in a container that can be overwintered under glass.
☐ Min. 10°C
E. grusonii, see p.37

Elaeagnus
Deciduous *E. angustifolia* makes a shrub or small tree, about 6m tall, with narrow silver leaves and tiny clove-scented flowers in summer. 'Quicksilver' ♥ has especially silvery leaves. Evergreen

CYTISUS × *PRAECOX*

ERIOGONUM ARBORESCENS

E. macrophylla tolerates shade and wind. It too bears fragrant flowers, and grows up to 3m. *E.* × *ebbingei*, also evergreen and slightly taller, makes a good screen.
☐ ✳✳✳

Eriogonum
Small range of shrubs, mostly tender. *E. arborescens* makes a rounded plant up to 1.5m. Tufts of leaves are borne at the shoot tips, along with white-pink flowers from summer to autumn. In cooler climates *E. umbellatum* var. *torreyanum* makes a hardy spreading shrub, up to 30cm high and 1m across, with bright yellow flowers. It looks striking in gravel.
☐ Min. 5°C/✳✳✳

Escallonia
Evergreen shrubs with glossy, dark green leaves. White, pink or red flowers are borne in summer. A good seaside shrub, resistant to wind and strong sunshine. Clip after flowering to keep in shape. Excellent for hedges and shelter belts. Height varies but is often around 1.5–2m.
☐ ✳✳

EUONYMUS FORTUNEI 'SILVER QUEEN'

Euonymus fortunei

Evergreen foliage shrub, about 60cm high, with leaves often splashed with gold or white. 'Silver Queen' ♀ is taller – about 2.5m – and has white-margined, dark green leaves. 'Emerald Gaiety' ♀, has similar colouring but grows to about 1m. 'Emerald 'n' Gold' ♀ has yellow-edged leaves.
❑ ✱✱✱

Euphorbia characias

Shrubby member of the euphorbia clan (*see p.60*). Metre-high stems of blue-green leaves carry brilliantly contrasting heads of green-yellow flowers from late spring. Cut faded flower stems back to the base. Sticky, milky sap can irritate skin.
❑ ✱✱

E. characias, see p.46

Fremontodendron californicum

A vigorous evergreen shrub with striking yellow flowers from late spring to autumn. *F.* 'California Glory' ♀ is a hardier, even more spectacular hybrid with deeper yellow flowers and a more spreading habit. Plant against a sunny, sheltered wall in temperate climates. Will reach about 6m high. Foliage can irritate skin.
❑ ✱✱/✱✱✱

Genista (Broom)

Range of mostly deciduous shrubs with yellow pea flowers, very similar to *Cytisus*. *G. hispanica* (Spanish gorse) forms dense, prickly mounds, about 75cm high, with flower spikes at the shoot tips in late spring and early summer. *G. aetnensis* ♀ (Mount Etna broom) is a

× *HALIMIOCISTUS WINTONENSIS* 'MERRIST WOOD CREAM'

large shrub or small tree, which can grow to 8m high, with elegant, weeping branches and fragrant flowers in mid- to late summer.
❑ ✱✱✱/✱✱

Genista aetnensis, see p.6

Griselinia littoralis ♀

Evergreen shrub with glossy, pale green leaves. Grow in mild seaside gardens as a hedge or as part of a shelter belt. Height varies according to growing conditions. Clip hedges in summer with secateurs. Leaves can have cream variegation.
❑ ✱✱/✱✱✱

× Halimiocistus wintonensis ♀

Spreading shrub, a cross between *Cistus* and *Halimium*, with green or grey-green leaves and saucer-shaped white flowers with dark crimson bands; 'Merrist Wood Cream' ♀ has creamy-yellow, red-banded blooms. Low growing, about 60cm, and excellent in raised beds or at the front of borders.
❑ ✱✱✱

EUPHORBIA CHARACIAS SUBSP. *WULFENII*

Halimium

A smaller version of × *Halimiocistus*, with the same kind of saucer-shaped flowers. It is especially suitable for containers on hot, sunny patios. Reaches 45cm high with a slightly wider spread.

❏ **

Hebe

The range of hebes includes several that make good ground-cover or container shrubs. *H. ochracea* is small and neat, growing to 45cm. Its dense stems covered with tiny leaves resemble whipcord. On 'James Stirling' ♀ these are ochre-yellow. Small white flowers appear in spring and early summer. *H. albicans* ♀ makes a 60cm, evergreen mound of grey-green leaves, with white flowers in early summer. The leaves of slightly smaller *H.* 'Red Edge' lose their red veining as its lilac flowers appear. *H.* 'Youngii' makes an evergreen, 25cm hummock with violet flowers in mid-summer. The glaucous foliage of *H. pinguifolia*

HEBE OCHRACEA
'JAMES STIRLING'

HEDERA COLCHICA 'DENTATA'

'Pagei' ♀ makes evergreen ground cover, 30cm high, with white flowers from mid-summer .

❏ **/***

H. 'Red Edge', *see p.40*

Hedera (Ivy)

Useful in shade, these easy evergreen climbers can also be grown as ground cover, the variegated types adding brightness to dull areas. *H. colchica* ♀ has large, leathery, dark green leaves and can grow to 10m. 'Dentata' ♀ has bright green leaves and stems that are flushed purple. The smaller-leaved common or English ivy, *H. helix*, has many variations in leaf shape and colouring, including some that are slightly tender.

❏ ▣ **/***

Helianthemum (Rock rose, sun rose)

Small, low, spreading, usually evergreen shrubs. *H. apenninum* has white flowers with bright yellow centres from spring to mid-summer. There are also numerous hybrids with pink, yellow, orange or scarlet flowers. Versatile plants for border fronts, rock gardens and raised beds. They also make good ground cover for sunny slopes.

❏ ***

Hippophae rhamnoides ♀
(Sea buckthorn)

A deciduous large shrub or small tree, up to 6m tall, with spiny shoots carrying narrow, grey-green leaves. Small yellow flowers appear in spring; plant both male and female plants to ensure orange berries. Excellent for hedges or windbreaks in coastal areas and for stabilizing sand dunes.

❏ ***

Hydrangea paniculata

A large deciduous shrub, about 3m high. Stems are topped by beautiful conical heads of creamy white flowers in late summer and early autumn. For the largest flowerheads, cut the previous season's shoots back to a woody framework in spring.

❏ ***

HIPPOPHAE
RHAMNOIDES

HYSSOPUS OFFICINALIS

Hypericum calycinum (Rose of Sharon)
A dwarf, semi-evergreen shrub, about 60cm high, with bright yellow flowers from mid-summer to mid-autumn. It spreads by runners and is a good ground-cover plant for dry banks. Invasive, but useful in that it tolerates shade.
▢ ▨ ✻✻✻

Hyssopus officinalis (Hyssop)
A dwarf, semi-evergreen, aromatic shrub bearing slender spikes of deep blue, occasionally pink or white, flowers from late summer to early autumn. Grows to about 60cm. Hyssop has a variety of herbal uses as well as being decorative.
▢ ✻✻✻

Juniperus (Juniper)
A range of evergreen conifers to suit just about any site. *J. procumbens* is one of the most successful for small, dry gardens, particularly on sandy soil or in windswept areas. It hugs the ground, reaching only 75cm high but spreading to about 2m, and has yellow-green needles. Slightly smaller *J. × pfitzeriana* 'Pfitzeriana Aurea' looks golden in all but the depths of winter when the leaves turn slightly greener.
▢ ✻✻✻

Lantana
Tender evergreen shrubs that can be used as ground cover in warm, frost-free climates. In cooler regions grow in a pot and take under cover in winter. *L. camara* has rounded heads of small flowers in a wide range of eye-catching colours, often combining two or more shades. *L. montevidensis* forms dense mats of foliage, up to 1m high, with lilac-pink to violet flowers in summer.
▢ Min. 10°C
L. montevidensis, see p.36

Lavandula (Lavender)
Intensely aromatic, small to medium evergreen shrubs (up to 1m tall) with fragrant, pale to deep purple flowers in mid- to late summer. Flowers can also be pink or white, and in some plants, the narrow leaves are particularly silvery. *L. angustifolia* is hardy, but some lavenders are less so. *L. stoechas* ♀ (French lavender) is borderline hardy (it will survive best in well-drained soil) and has curious dark purple flowers topped by purple bracts; it grows to 60cm. Use lavender for edging or as a low hedge. Trim plants in spring to prevent them from getting leggy, but avoid cutting into old wood.
▢ ✻✻/✻✻✻
L. angustifolia 'Twickel Purple' ♀, *see p.18*

Lavatera
Range of shrubs (as well as annuals, biennials, perennials) often found growing in the wild in dry, rocky places. 'Barnsley' ♀ is vigorous and semi-evergreen, reaching 2m, with grey-green leaves and a graceful habit. White flowers, with a deep pink eye, appear in profusion from mid-summer and slowly turn soft pink. Plants soon become leggy and are best pruned while young. They can also be cut back by frost but usually re-shoot.
▢ ✻✻✻

LAVATERA 'BARNSLEY'

Lonicera periclymenum

(Common honeysuckle)
Beautifully fragrant white to yellow flowers, often flushed red, appear in mid- to late summer. A deciduous, twining climber that can reach 7m. The flowers of 'Serotina' ♀ (late Dutch) are streaked with reddish-purple. 'Graham Thomas' ♀ has white flowers, turning yellow. Cut hard back after flowering to produce new shoots near the base.
◘ ◙ ✻✻✻

Lotus hirsutus

Attractive small, shrubby plant, about 60cm high. Evergreen or semi-evergreen grey-green leaves are softly coated with silver hairs. The pea-like, pinkish-white flowers in summer and early autumn develop into reddish-brown seed pods. May not survive a wet, cold winter.
◘ ✻✻/✻✻✻

Nerium oleander

Leggy shrub or small tree, about 2m high or more in favourable climates, with narrow green or grey-green

LONICERA PERICLYMENUM
'SEROTINA'

OPUNTIA ROBUSTA

leaves. A profusion of red, pink or white flowers appear throughout summer. Although it can withstand drought, it is susceptible to wind damage, and should not be allowed to become leggy. Needs a protected spot in cool climates and tolerates some shade. All parts are toxic.
◘ ✻

Olearia

Mainly evergreen shrubs or small trees with white daisy flowers. O. × haastii and O. macrodonta ♀ are suitable for hedging and shelter belts in windswept and seaside areas. Trim in mid- to late spring. O. × haastii (up to 2m) flowers from mid-summer; its glossy leaves are felted underneath. O. macrodonta, up to 6m tall, has holly-like leaves and flowers in summer. O. × scilloniensis, also 2m high, flowers in late spring.
◘ ✻✻/✻✻✻

Opuntia robusta

One of the many prickly pear cacti – architectural and dramatic when it takes on a tree-like shape at about 2m.

Oval-shaped stem sections are covered in tufts of spines. Yellow flowers in summer are followed by red fruits. It sometimes survives outside in temperate regions in sheltered spots with a favourable microclimate, free of frost and damp. Alternatively, grow in a container and overwinter under glass. Slow-growing.
◘ Min. 7–10°C
O. robusta, see p.37

Ozothamnus rosmarinifolius

A compact, erect shrub, usually 2–3m high, with needle-like leaves. Fragrant white flowers emerge from red buds in early summer. Avoid planting in heavy soils that lie wet in winter.
◘ ✻✻

Parahebe catarractae ♀

An evergreen shrubby plant with small leaves tinged purple when young and flowers throughout summer. About 30cm high and wide, it makes a fine plant for gravel gardens and border fronts.
◘ ✻✻✻

PARAHEBE
CATARRACTAE

PEROVSKIA 'BLUE SPIRE'

Perovskia
Upright, deciduous shrub, with striking, grey-white stems and silver-grey leaves. Long spikes of violet-blue flowers appear in late summer and early autumn. Usually reaches about 1.2m. Should re-shoot if cut back by frost. Useful in containers, on dry, chalky soil and by the coast.
☐ ✳✳✳
P. 'Hybrida', *see p.19*

Phlomis
The range includes shrubs and herbaceous perennials. *P. fruticosa* ♀ (Jerusalem sage) is a rounded, evergreen shrub, about a metre high, with yellow flowers in early and mid-summer among grey-green leaves. An architectural plant, useful for containers. *P. italica* has lilac-pink flowers and silvery, felted leaves. (*See also P. russeliana, p.62.*)
☐ ✳✳/✳✳✳

Prunus laurocerasus ♀
(Cherry laurel)
Thick, glossy evergreen shrub, 8m tall and wide but spreading more with age. Spikes of small, fragrant white flowers

ROMNEYA COULTERI 'WHITE CLOUD'

are produced in mid- to late spring, followed by cherry-like red fruits that ripen to black. Makes a dense hedge as it withstands hard pruning, preferably with secateurs.
☐ ✳✳✳

Romneya coulteri ♀
(Tree poppy)
Large, papery, white flowers with bright yellow centres are produced nearly all summer above attractive grey-green foliage. Deciduous, and needs free-draining soil plus the protection of a sheltered wall in frosty areas. May be difficult to establish, but when growing well can sucker extensively and may become invasive. Will reach 1–2.5m high.
☐ ✳✳

Rosa (Rose)
The majority of roses need reasonably moist conditions but a few of the species cope with dry conditions and salty winds. *R. rugosa* is a vigorous rose with wrinkled, leathery leaves. Grows 1–2m tall and makes a splendid, intruder-

proof hedge. Single, scented, violet-carmine flowers are followed by showy red hips. *R. rugosa* var. *rosea* has pink flowers. *R. pimpinellifolia* (burnet rose) has very small, fern-like leaves, lots of prickles and single, creamy-white flowers followed by black hips. A spreading habit and it grows to about 1m tall.
☐ ✳✳✳

Rosmarinus officinalis
(Rosemary)
Evergreen shrub with superbly aromatic, needle-like leaves. Tubular blue flowers cover the stems in spring; there may be a second flush in autumn. Will grow to 1.5m tall, but can get straggly; prune if necessary after flowering. Can be grown as a hedge; clip after flowering. Trailing or prostrate forms grow to only 15cm high – excellent for banks and raised beds but less hardy.
☐ ✳✳

Ruta graveolens (Rue)
Evergreen shrub grown for its filigree, pungent, blue-green foliage; yellow flowers appear

ROSA RUGOSA

SANTOLINA
CHAMAECYPARISSUS

TAMARIX RAMOSISSIMA
'PINK CASCADE'

in summer. Grows to a metre high. Clip off flower buds for the best foliage. Contact with foliage in sun can cause skin to blister.
❏ ✳✳✳

Santolina (Cotton lavender)
Compact, evergreen, rounded shrub covered with bright yellow or lemon button-like flowers in mid- to late summer. Foliage is better if the flowers are removed before they open. *S. chamaecyparissus* ♀ grows to about 50cm high and has white-woolly shoots and finely divided, aromatic, silvery leaves. Suitable for a low hedge. Clip in late spring. *S. pinnata* is slightly larger and may have green or grey-green leaves. Prune in spring to prevent it from sprawling; avoid cutting into old wood.
❏ ✳✳
S. pinnata 'Edward Bowles', see p.18

Spartium junceum ♀
(Spanish broom)
An upright shrub, up to 3m tall. The thin, sparsely leaved stems bear fragrant yellow

pea flowers from summer to early autumn. Flattened brown seed pods follow the flowers. Thrives by the coast and on chalky soil. In frost-prone areas, grow in the shelter of a sunny wall. Trim lightly in spring, if necessary.
❏ ✳✳

Tamarix (Tamarisk)
Graceful shrubs or small trees, from 3m to 5m tall, with arching stems and delicate feathery foliage. Plumes of small pink flowers appear on *T. ramosissima* in late summer and early autumn; *T. tetrandra* ♀ flowers in spring. Plants grow well on sandy soil and can be used to make an excellent hedge or windbreak in exposed coastal regions.
❏ ✳✳

Teucrium fruticans
Evergreen shrub with an open habit, growing to a metre tall. Small, silvery, aromatic leaves clothe branching stems, and lipped, pale blue flowers appear in summer. Needs well-drained soil and usually

the shelter of a sunny wall. In warm climates, can be grown as a hedge. Clip in spring.
❏ ✳✳

Ulex europaeus (Furze, gorse, whin)
Dense, bushy shrub, up to 2.5m, with viciously spiny leaves and stem tips. Bright yellow pea flowers are produced intermittently all year, but are at their peak in spring. A plant that likes poor, sandy, acid to neutral soil, and can be invasive. Use to form an impenetrable low hedge, and clip after flowering every other year.
❏ ✳✳✳

Yucca
Dramatic and architectural with evergreen sword-like leaves and spikes of white bells from late summer. Leaves may be striped yellow or cream. *Y. filamentosa* ♀ is fully hardy and reaches 75cm; *Y. gloriosa* ♀ reaches 2m but is less hardy. *Y. whipplei*, reaching about 1m, has slender grey-green leaves.
❏ ✳✳/✳✳✳
Y. whipplei, see p.36

TEUCRIUM FRUTICANS 'COMPACTUM'

PERENNIALS AND BULBS

ROUGHT-TOLERANT PERENNIALS seldom produce the same sort of luxuriant growth as traditional herbaceous border plants, so require minimal staking and cutting back, other than to deadhead. A few need all-year warmth; in cool climates overwinter them in a greenhouse, conservatory or cold-frame. Many rock garden plants (*see p.66*) are also ideal for border edges and containers.

Acanthus
Handsome plants with large, divided leaves. Spires of hooded, tubular flowers in white, green, dusky purple or pink appear on stems up to 1.2m tall. Summer-flowering *A. spinosus* ♀ has very deeply cut leaves in clumps up to 1.5m across.
❑ ✳✳✳

Achillea
Spreading or clump-forming range of perennials with attractive fern-like, grey or green leaves. Long-lasting, large, flat flowerheads, in shades of yellow or palest peach to deep red, are carried on stout stems from early summer to early autumn. Some, such as 'Moonshine' ♀ and *A. filipendulina* 'Gold

ACHILLEA FILIPENDULINA 'GOLD PLATE'

Plate' ♀, are evergreen; heights can vary from 60cm to 1.2m. Grows well on poor, dry soil. Flowers can be dried for winter decoration.
❑ ✳✳✳
A. millefolium 'Sammetriese', *see p.30*

Agapanthus
Stout stems crowned with round heads of blue or white flowers rise from clumps of strap-like leaves in summer. Headbourne hybrids are hardy but variable in colour; flower stems reach 80cm. 'Loch Hope' ♀ produces deep blue flowers in late summer on 1.5m stems. Much smaller 'Blue Moon' is good in pots.
❑ ✳✳/✳✳✳
A. 'Blue Moon', *see p.40*

Agave
Tender succulents with stiff fleshy leaves, sometimes variegated and usually tipped with a vicious spine. The greyish-green leaves of *A. americana* ♀ are up to 2m long, and have pale yellow margins in 'Marginata'. *A. filifera* ♀ has rosettes of slender leaves up to 25cm long. Plants can sometimes survive outside in cooler regions in mild sites, free from frost and damp. Or agaves can be grown in pots and overwintered under glass.
❑ Min. 10°C

Allium
Range of ornamental onions, usually with round heads of small starry flowers in pinks, mauves, yellow or white. Strappy leaves die back as the flowers appear. Most thrive in sandy soil. *A. caeruleum* ♀ has blue flowers on 60cm stems in early summer. *A. cristophii* ♀ is a similar height but with heads up to 20cm in diameter of metallic pink-purple flowers in early summer.
❑ ✳✳/✳✳✳
A. cristophii, *see p.31*

Aloe
A showy, tender succulent with rosettes of thick tapering leaves. *A. ferox* grows very slowly to 4m, and looks like a single-stemmed tree crowned by a rosette of glaucous,

ALLIUM CAERULEUM

Coronilla valentina - Glauca 'Citrina'

ots { ot Rose - Gertrude Jekyll. ✓

2 Hydrangea arborescens ✓'Annabelle'

Clematis - Etoile Rose

Kit. { Cistus Aguilarii - Maculatus. ✓
Side. { + Cotinus ?

Day room corner ?

2 Gaura ? Nepeta ?

ASCLEPIAS TUBEROSA

Asclepias tuberosa
(Butterfly weed)
Clusters of orange-red flowers, sometimes yellow, are carried on strong stems, up to 90cm tall, from mid-summer to early autumn. The flowers, attractive to butterflies and bees, are followed by spindle-shaped green fruits.
☐ ✳✳✳

Asphodeline lutea
(Yellow asphodel)
Stiff spikes of yellow, starry flowers up to 1.5m tall emerge from clumps of blue-grey, grassy leaves. Stems become beaded with green cherry-like fruits, which later turn brown. The leaves tend to die down and disappear after mid-summer and re-shoot in early autumn.
☐ ✳✳✳
A. lutea, see pp.28, 31

Calamintha nepeta
A tough ground-cover plant, also good for gravel gardens. Grows up to 45cm. Leaves smell minty and the profusion of tiny lilac flowers from summer to autumn attract

DICTAMNUS ALBUS VAR. PURPUREA

CRAMBE MARITIMA

bees. The flowers of 'Blue Cloud' are a good deep blue.
☐ ✳✳✳
C. nepeta, see p.19

Crambe
Clump-forming perennials that bear 2.5m-high clouds of tiny white flowers. The mounds of large, puckered leaves produced by *C. cordifolia* ♀ die down in mid-summer as·the flower stems emerge. Seakale, *C. maritima*, has glaucous leaves; shoots can be blanched in late winter, steamed and eaten as a vegetable.
☐ ✳✳✳
C. maritima, see pp.16, 18

Dictamnus albus ♀
(Burning bush)
Pretty spires of white or pale pink flowers are produced in early summer above clumps of divided leaves. Height can vary from 40cm to 90cm. In hot, still weather, a flammable vapour is produced. This can be ignited near the leaves without damaging the plant. Contact with skin may cause photodermatitis.
☐ ✳✳✳

spiny leaves. *A. aristata* ♀ has dense rosettes of leaves and 60cm-high spires of orange-red flowers. Like agaves (*opposite*) aloes can be grown outside in mild, exceptionally well-drained sites, or in pots and protected in winter.
☐ Min. 10°C
A. ferox, see p.37

Anthemis
Perennials (and some annuals) forming attractive clumps of parsley-like aromatic foliage. White or yellow daisy flowers have a long season from late spring or early summer. *A. punctata* subsp. *cupaniana* ♀ has fine feathery, grey foliage.
☐ ✳✳/✳✳✳
A. tinctoria 'E.C. Buxton', *see p.6*

Artemisia
Invaluable range of perennials and shrubs (*see p.50*) for hot sites. Many make good ground cover, their silver foliage setting off other plants. Low-growing *A. stelleriana* resists drying winds and strong sun. Evergreen, it reaches 15cm high, spreading to 45cm.
☐ ✳✳✳

ERYNGIUM ALPINUM

Echinacea purpurea
(Purple cone flower)
An easy-to-grow plant that
adds to the colour of the late
summer garden. Distinctive
mauve-pink or white daisies
have high central domes of
orange-brown.
◪ ✻✻✻
E. purpurea 'Robert Bloom',
see p.31

Echinops (Globe thistle)
Spiny-leaved, clump-forming
plants with round, metallic-
blue flowerheads in summer.
E. bannaticus may reach 1m;
E. ritro is shorter (up to 60cm)
with flowers ageing to a
darker blue. 'Veitch's Blue' has
a slightly longer flowering
season. Flowers dry well for
indoor arrangements, or look
handsome left on the plant.
◪ ✻✻✻
E. ritro, *see p.30*

Epimedium (Barrenwort)
Invaluable ground-cover
plants that tolerate dry shade.
Can be evergreen or
deciduous, up to 25cm tall,
with glossy leaves sometimes
bronze-tinted in spring and
autumn. Dainty flowers, often

spurred, appear from early
spring to early summer in
yellow, beige, white, pink, red
or purple. *E. × versicolor* is
one of the most tolerant of
dry soil and sun. Shear over
early in spring to promote
fresh growth.
◪ ✻✻✻

Eryngium (Sea holly)
Wide range of perennials (also
some annuals and biennials),
usually with deep tap roots
and basal rosettes of spiny,
veined leaves. In summer,
round or egg-shaped, white or
blue flowerheads are held
above attractive ruffs of spiky
bracts. *E. giganteum* ♀ is
short-lived and often grown as
a biennial. It reaches about
90cm and self-seeds liberally.
E. alpinum ♀ is shorter with
very handsome flower ruffs.
E. × oliverianum ♀, a cross
between these two, has well-
marked veining. Leave faded
flowerheads for winter display.
◪ ✻✻✻
E. giganteum, *see p.18*
E. maritimum, *see p.33*

Erysimum (Wallflowers)
Evergreens with narrow grey-
green leaves and spikes of
mauve, orange, red, yellow or
cream flowers from late
winter to early summer. Woody
stems can get very leggy and
plants need annual renewing.
Plants reach 25–75cm.
◪ ✻✻✻

Euphorbia (Spurge)
Diverse range of plants, often
with striking blue-green
foliage and contrasting
yellow-green flowerheads.
Drought-tolerant perennials
include *E. myrsinites* ♀ with
succulent, pointed leaves on

almost prostrate stems,
arranged like spiders' legs.
Bright greenish-yellow flowers
appear at the stem tips in
spring. *E. nicaeensis* has
similar foliage but is bushy
and upright, with flowers
from spring to mid-summer.
E. griffithii, about 75cm high,
has orange-red flowerheads in
early summer and leaves that
take on red and yellow tints
in autumn. A spreading plant,
it can be invasive. Milky sap
can cause skin irritation. (*See*
also *E. characias, p.52.*)
◪ ✻✻✻
E. myrsinites, *see p.19*
E. griffithii, *see pp.28, 31*

Gaillardia
Bushy perennials, often short-
lived, found growing on the
prairies of North America.
Colourful flowers are set off
by the grey-green leaves. Easy
to grow. *G. × grandiflora*
reaches about 90cm and
produces yellow daisy flowers,
banded with red, from early
summer to early autumn.
'Dazzler' ♀ has orange-red
flowers with yellow tips and a
maroon centre.
◪ ✻✻✻

GAILLARDIA × GRANDIFLORA

GAZANIA CHANSONETTE SERIES

Gazania

Bright daisy flowers are produced all summer but only open when the sun is out. Spreading plants are about 20cm high. Though perennial, best treated as an annual in temperate climates. Evergreen leaves are covered with silky-white hairs underneath. The flowers come in shades of bronze-orange, rose, salmon, orange and yellow, often banded with green. Excellent plants for containers.
◨ *
Harlequin Hybrids, *see p.41*

Helichrysum

Wide range of perennials, some grown purely for their attractive woolly or hairy foliage, others also for their clusters of small, papery flowers. *H. petiolare* ♥ has long, trailing stems of evergreen grey-green leaves, especially useful in containers but best treated as an annual in temperate climates. *H.* 'Schwefellicht' forms silvery clumps of foliage with sulphur-yellow flowers in late summer, up to 40cm tall.
◨ ✳✳✳

Iris

Several types of iris enjoy dry conditions, including the bearded irises. They come in a range of sizes – dwarf, intermediate and tall. Irises offer a wide choice of colour in their flowers, some with ruffled petals. The handsome spears of foliage grow out of surface-rooting rhizomes that like to bake in the sun and need regular division in late summer.
◨ ✳✳✳

Kniphofia (Red hot poker)

Striking flower spikes in lemon, orange or red shoot up in summer from clumps of strap-like leaves. Versatile plants, able to grow well in a wide range of conditions, but needing plenty of organic matter adding to poor, dry soil to help them survive long periods of drought.
◨ ✳✳✳

Lamium

Spreading plants, good for low ground cover in dry shade (where they are less invasive than in damp conditions). *L. maculatum* grows to 20cm with a 1m spread, and has pink flowers in summer. 'White Nancy' ♥ is a similar size with white flowers and silvery leaves edged green.
◨ ✳✳✳
L. maculatum, see p.34

Lampranthus

Low-growing succulents, covered in summer with bright daisy flowers. Native to the semi-desert areas of South Africa, they are ideal in arid-landscape gardens but must be overwintered under glass in temperate climates.

L. haworthii has large pink-purple flowers all summer, and pale green foliage, frosted with grey. Grows to 50cm high with a wider spread. *L. spectabilis* is slightly lower growing and is available in a range of apricots and reds.
◨ Min. 7°C
L. haworthii, see p.36

Limonium platyphyllum (Sea lavender)

Rosettes of spoon-shaped leaves are topped, in late summer, by masses of tiny lavender-blue flowers, useful in dried flower arrangements. These are borne on 60cm-tall, branching, wiry stems. A good plant for dry coastal areas and sandy soil.
◨ ✳✳✳

Lychnis

Lychnis originate from a variety of habitats; those with the most silvery foliage generally tolerate the driest conditions. *L. flos-jovis* forms mats of greyish leaves with white, pink or scarlet flowers, 20–60cm high, from early to late summer.
◨ ✳✳✳

LIMONIUM LATIFOLIUM

NEPETA SIBIRICA

Nepeta (Catmint)
An extremely useful range of perennials, usually blue-flowered and with aromatic leaves. *N. sibirica* is well able to withstand drought. It grows to 90cm, and bears lavender-blue flowers above dark green leaves from mid- to late summer. *N. × faassenii* has seemingly never-ending sprays of lavender-blue flowers, at their best in mid-summer, above 45cm-high clumps of silvery-grey leaves. Shear back after flowering to keep plants neat and to encourage more flowers.
◫ ✳✳✳
N. × faassenii, see p.31

Oenothera (Evening primrose)
Yellow, or occasionally white or pink, cup-shaped flowers open at dawn or dusk, fade quickly but usually appear over a long period. Most perennial oenotheras grow well in sunny, well-drained sites. *O. macrocarpa* ♀ has trailing, red-tinted stems, 15cm high but spreading to 50cm, with yellow flowers from late spring to autumn.
◫ ✳✳✳

Origanum, see p.67

Ornithogalum umbellatum (Star of Bethelehem)
Perennial bulb with open clusters of star-shaped white flowers, striped green on the outside, in early summer. Silver-striped leaves fade as the flowers open. Can reach 30cm high.
◫ ✳✳✳

Pachysandra terminalis ♀
Low, spreading evergreen, up to 20cm high, with glossy, toothed leaves clustered at the ends of the stems and tiny white flowers in early summer. An excellent plant for dry soil in shade, especially under trees.
◪ ✳✳✳

Phlomis russeliana ♀
Produces weed-smothering clumps of large hairy leaves. Stout stems, up to 90cm tall, bear cream and butter-yellow hooded flowers that open from mid- to late summer. If left on the plants, the brown seedheads will decorate the garden in winter.
◫ ✳✳✳

OENOTHERA MACROCARPA

PHLOMIS RUSSELIANA

Phormium (New Zealand flax)
Dramatic evergreen plants with sword-shaped leaves. Several colour variations available including all-bronze or copper- or yellow-striped leaves. *P. tenax* ♀, the hardiest, forms clumps of 3m-long leaves with taller spikes of dull red flowers in summer. Although of borderline hardiness, plants are more likely to survive winter cold if the roots are in light, well-drained soil. You can also wrap leaves in a sheath of hessian or bubble-wrap to protect them further.
◫ ✳✳/✳✳✳

Salvia
Herbaceous salvias require a variety of conditions; the following suit sunny, dry sites. From mid- to late summer, *S. × superba* ♀ produces 80cm-tall spikes of violet flowers with red bracts that persist after the petals have fallen. *S. argentea* ♀ could be grown purely for its furry, tactile leaves, which form large silver rosettes. Good on gravel but needs protection

from winter wet. (*See also S. sclarea, p.69.*)

🔲 ✳✳✳

Saponaria

Perennials and annuals for dry soils, varying in height from 5cm to 75cm. *S. officinalis* (soapwort) grows to 60cm high, with double pink, red or white flowers in summer and autumn. Those such as pink-flowered *S. × olivana* ♧, which form spreading mats of foliage, are good for troughs, dry walls and rock gardens.

🔲 ✳✳✳

Sedum

Clumps of fleshy stems with slightly scalloped, grey-green leaves are topped in late summer with flattish heads of star-shaped flowers. Most sedums attract butterflies and bees. At flowering time, plants can have a tendency to sprawl. *S.* 'Ruby Glow' ♧ has dusky red flowers and grows to 25cm high. *S.* 'Bertram Anderson' ♧ is a similar height and has purple foliage and flowers. *S. spectabile* ♧ (ice plant) has pink flowers on

TULIPA TARDA

VERBASCUM OLYMPICUM

45cm stems. There are also several small sedums grown as rock plants (*see p.67*).

🔲 ✳✳✳

S. 'Bertram Anderson', *see p.18*; *S.* 'Ruby Glow', *see p.40*

Stachys

The leaves of *S. byzantina* (lambs' lugs) form rosettes of soft, woolly grey; spikes of pink flowers appear in mid-summer. 'Silver Carpet' is a fine non-flowering silver carpet plant. *S. candida* is a small, spreading, shrubby plant, up to 15cm high, with rounded, felted, grey-green leaves. White flowers streaked with purple are produced in summer. Excellent for gravel gardens and is best not exposed to winter wet.

🔲 ✳✳✳

Tulipa

For many tulips hot, dry summers are essential. These include low-growing *T. tarda* ♧, *T. greigii* and *T. kaufmanniana*. The Darwin hybrids and other taller bedding tulips also cope well with dry conditions.

🔲 ✳✳✳

Verbascum

Wide range of perennials (plus a few annuals and biennials), with spikes of yellow, white or pink flowers that shoot up from basal rosettes of leaves. These are often grey-green. *V. olympicum* has woolly, whitish leaves with imposing 2m-high candelabras of yellow flowers in late summer. It may die after flowering. *V. chaixii* 'Pink Domino' ♧ has dark, purplish-green leaves and 70cm spikes of rose-pink flowers from early to late summer.

🔲 ✳✳✳

Verbena bonariensis

An extremely useful plant for late summer and autumn, when clusters of small rosy-purple flowers appear on wiry, branching stems, about 2m tall. Can be used to create flowering screens through which you can see other parts of the garden. It is not reliably hardy but often self-seeds itself among other plants in an attractive, serendipitous way.

🔲 ✳✳

V. bonariensis, *see p.18*

VERBENA BONARIENSIS

ORNAMENTAL GRASSES

NOT ALL GRASSES ARE DROUGHT TOLERANT – a few like it damp – but those that suit dry conditions deserve a place in the garden for the way they bend in the wind, catch the sunlight, and look magnificent beaded with dew or rimed with frost. They come in a whole range of sizes; unless stated otherwise, those below are perennial, many of them evergreen, giving all year decoration.

Briza (Quaking grass)
Forms dense clumps of foliage with heart-shaped flowers that quake in the breeze.
B. media is perennial, with blue-green leaves and nodding flowers in summer, on 60cm stems; they turn straw-coloured as they mature.
◻ ✳✳✳

Cortaderia selloana
(Pampas grass)
A spectacular grass with tall plumes of silky flowers, often flushed pink or purple, in late summer. Flower stems can reach 3m high. 'Pumila' makes a smaller plant. The sharp-edged leaves form dense evergreen clumps. Plants look good in minimalist designs with decking and cobbles.
◻ ✳✳✳

CORTADERIA SELLOANA
'AUREOLINEATA'

Elymus hispidus
(Blue wheatgrass)
An evergreen perennial with narrow, intensely blue leaves that grow to about 60cm. These are erect at first, but spread out later. The wheat-like flower spikes are the same brilliant blue at first, but fade to beige. The blue-grey leaves of *E. magellanicus* have a whitish bloom.
◻ ✳✳✳

Eragrostis curvula
(Love grass)
A perennial that makes large, graceful clumps with arching, rough-textured, dark green leaves up to 1.2m tall. In late summer it forms a haze of nodding spikelets, which persist to give a metallic grey effect above wintry buff stems.
◻ ✳✳

Festuca (Fescue)
A perennial grass growing in dense clumps. The very fine blades range from green to intense steely-blue. *F. glauca* forms tidy mounds of powder-blue leaves up to 30cm high. Spikelets of violet-flushed, blue-green flowers appear in early and mid-summer. 'Blaufuchs' ♀ has bright blue leaves. 'Elijah Blue' has eye-catching steely-blue leaves, to 20cm, and blue-grey flowers.
◻ ✳✳✳
F. ovina, see p.33

Helictotrichon sempervirens ♀
(Blue oat grass)
A densely tufted grass that forms a mound of narrow, grey-blue leaves. Evergreen, though it can look rather tattered in winter. Spikelets of straw-coloured flowers, marked with purple, are borne on stiff, 1.4m stems in early and mid-summer.
◻ ✳✳

Holcus mollis 'Albovariegatus'
A creeping plant with tufts of flat, blue-green and cream leaves, up to 20cm tall. It gives the overall effect of a carpet of white, especially in spring. Spikes of pale green flowers are borne in summer.
◻ ◼ ✳✳✳

FESTUCA GLAUCA

HELICTOTRICHON SEMPERVIRENS

Koeleria glauca
(Glaucous hair grass)
Evergreen tufts of narrow, silver-blue leaves with inrolled margins reach about 20cm high. Shiny spikelets of cream and green flowers in early and mid-summer gradually turn buff-coloured with age.
◻ ✳✳✳

Leymus arenarius
(Lyme grass)
Although fast-spreading in dry gardens, sometimes invasively so, it is desirable where it can be contained or given room for its broad blue-green leaves. They grow to 60cm high. Tall, stiff, wheat-like flower spikes are borne throughout summer.
◻ ✳✳✳

Melica altissima
A dainty, erect, non-invasive perennial making soft-leaved clumps. 'Atropurpurea' produces metre-tall spikes of deep purple flowers in mid-summer, which fall charmingly to one side and fade to rosy-pink when dried.
◻ ✳✳✳

Pennisetum (Fountain grass)
A fountain effect is created by the way the fluffy, fox-tail flowerheads rise up on slender stems from nicely rounded mounds of green leaves. *P. alopecuroides* is perennial and evergreen, making graceful clumps about 75cm high of arching, bright green leaves. The brownish-purple bottlebrush flowers rise about a metre above the leaves in autumn and early winter, supplying the garden with marvellous colour. They pale with age but last through the winter. 'Woodside' is very free-flowering. The dwarf 'Hameln', only 30cm high, has dark green leaves that turn golden yellow in autumn. It flowers freely, but its stiffish stems lessen the "fountain" effect. They are all slightly tender. *P. villosum* has very soft, fluffy flowers that emerge silky green then turn light pink in colour. The narrow green leaves are about 15cm long. It is often grown as an annual in cold areas.
◻ ✳✳

PENNISETUM VILLOSUM

STIPA GIGANTEA

Schizachyrium scoparium
(Little blue stem)
Clumps of erect mid- to grey-green leaves turn a purplish orange-red in autumn. Wispy metre-tall flower spikelets appear from mid-summer to late autumn.
◻ ✳✳✳
S. scoparium, see p.19

Stipa
Decorative grasses with beautiful, often feathery flowerheads. *S. tenuissima* makes dense tufts of bright green deciduous leaves. The wispy flowers, up to 60cm tall, are produced all summer. These gradually turn a buff colour and billow in the gentlest puff of wind. The dark green leathery leaves of *S. arundinacea* turn orange-brown in autumn, enlivening the winter garden. Arching stems, up to 1m high, of purplish-green flowers appear from mid-summer. Evergreen or semi-evergreen *S. gigantea* ♀ has 2m stems with large, open heads of oat flowers that tremble and glisten in the sun.
◻ ✳✳/✳✳✳
S. arundinacea, see p.30

ROCK PLANTS

MOST ROCK PLANTS ARE IDEALLY SUITED to the dry garden. In their native habitats they need to survive long periods of drought, intense light and scant soil, though they do often naturally tuck themselves behind stones or boulders that trap all available moisture and cast a little shade. Many make good candidates for containers and for planting among paving stones or edging paths.

Acaena

A. novae-zelandiae is a mat-forming evergreen, no higher than 15cm, ideal for covering banks in full sun. Filigree leaves can be green to grey-green. Red burrs follow mid- to late-summer flowers. *A. saccaticupula* 'Blue Haze' is vigorous with dark red burrs and grey-blue leaves.
❂ ✻✻✻

Alyssum

A good plant for border fronts and rock and scree gardens. Both *A. montanum* and *A. wulfenianum* are spreading evergreens, reaching about 15cm high, with grey-green leaves. Dense heads of tiny yellow or lemon flowers respectively cover the plants in early summer.
❂ ✻✻✻

ALYSSUM WULFENIANUM

ANTENNARIA DIOICA 'ROSEA'

Antennaria dioica

Forms flat mats, up to 45cm wide, of semi-evergreen, silvery leaves topped by fluffy, white or pink "everlasting" flowers in late spring and early summer. Makes good ground cover and can also be grown in wall crevices and paving. The flowerheads may be dried for decoration.
❂ ✻✻✻

Arabis

Easily grown evergreens with clusters of small white or purple flowers in late spring. *A. caucasica* forms low mats, up to 50cm wide, of grey-green leaves with white flowers. Good for ground cover and will spread rapidly over dry walls. 'Variegata' has pale yellow leaf margins.
❂ ✻✻✻

Armeria (Sea pink, thrift)

Eye-catching small, round heads of white or pink-purple flowers are borne in profusion in late spring and early summer above hummocks of grass-like leaves. *A. maritima*, found growing wild along sea cliffs, grows to 20cm tall. *A.* 'Bee's Ruby' ♥ has large, bright pink flowers on slightly taller stems. Excellent for gravel gardens.
❂ ✻✻✻

Aubrieta

Low, carpeting plants, spreading to 60cm or more, that tumble over walls and rocks and grow well with very little moisture. Profuse, single or double, white, pink, mauve or purple flowers are borne in spring. Tiny evergreen leaves can be attractively variegated. Cut back after flowering.
❂ ✻✻✻

Cerastium tomentosum

(Snow-in-summer)
Very vigorous, low, mat-forming plant with small, woolly, almost white leaves and a profusion of small white flowers in late spring and summer. Tough and will succeed in poor conditions. Cut back as necessary after flowering so that it does not scramble out of control.
❂ ✻✻✻

Gypsophila repens ♀

A spreading, semi-evergreen perennial, covered with dainty, white or pink flowers for long periods in summer. Grows to 20cm tall with a spread of up to 50cm. 'Dorothy Teacher' ♀ is more compact with blue-green leaves and pale pink flowers that darken with age.
⬚ ❋❋❋

Origanum

Range of decorative, aromatic plants related to marjoram, the culinary herb. Many have unusual, attractive flowers. *O. amanum* ♀ is a spreading evergreen, up to 20cm high, with pink flowers among green-pink bracts from late summer. *O. laevigatum* ♀ has upright, wiry stems, to 60cm, with sprays of tiny purple flowers from late spring to autumn. Evergreen *O.* 'Kent Beauty' has trailing stems tipped with pink flowers. There are also gold-leaved or variegated origanums. All suit borders, rock gardens and containers if protected from winter wet.
⬚ ❋❋❋
O. laevigatum, see p.31

ORIGANUM 'KENT BEAUTY'

PHLOX SUBULATA 'G.F. WILSON'

Phlox subulata

An evergreen, mat-forming phlox, spreading up to 50cm. Small purple, red, lilac, pink or white flowers appear in late spring and early summer. It grows well in dry sites and likes sun, unlike other trailing phloxes.
⬚ ❋❋❋

Rhodanthemum hosmariense ♀

White daisies appear non-stop from spring to autumn on a spreading plant, up to 30cm tall, with fine silvery foliage.
⬚ ❋❋

RHODANTHEMUM HOSMARIENSE

Saponaria, see p.63

Sedum

Apart from the sedums grown as herbaceous perennials (*see p.63*), there are several low-growing types suitable for rock gardens and troughs. *S. spathulifolium* 'Purpureum' ♀ forms 10cm-high mats of red-purple leaves with star-shaped yellow flowers in summer.
⬚ ❋❋❋
S. spathulifolium 'Purpureum', *see p.33*

Sempervivum (Houseleek)

Fleshy rosettes of green to purple leaves spread to form mats. They provide ground cover over really shallow soil, and even grow on walls and tiled roofs. Thick-stalked, reddish-pink flowers are produced in summer. The foliage of *S. arachnoideum* ♀ has a cobwebbing of white hairs; others have reddish tints to the foliage. All can withstand extreme drought.
⬚ ❋❋❋
S. ciliosum, see p.33
S. montanum, S. giuseppii, see pp.40–41

Thymus (Thyme)

Can form either spreading mats of foliage or small bushy plants, and the tiny aromatic leaves may be green, gold, or variegated yellow or silver. Flowers, in summer, are white, pink or purple and adored by bees. *T. serpyllum* is excellent creeping out from cracks between paving slabs or over banks and low walls. *T. vulgaris* 'Silver Posie' makes a short woody, bushy plant (about 20cm high) with pretty white-margined leaves.
⬚ ❋❋❋

ANNUALS AND BIENNIALS

THE MAJORITY OF ANNUALS flower all the better in fairly impoverished ground: too rich a soil encourages a lot of leaf. Hardy annuals can be sown in autumn to give them an early start the following year; many self-seed of their own accord – be strict about pulling out unwanted seedlings. Biennials produce flowers in their second season; they, too, often scatter their own seed.

Argemone (Prickly poppy)
The sprawling stems of *A. grandiflora* carry prickly, glaucous leaves and clusters of slightly scented white poppies in summer. Usually annual (sometimes perennial), it can grow to almost a metre tall. *A. mexicana* reaches the same sort of height. A clump-forming annual, it has blue-green leaves and yellow poppy flowers. Both thrive on poor sandy or stony soil.
◻ ✻

Brachyscome iberidifolia (Swan river daisy)
A bushy annual, about 45cm tall, with slender stems and soft, deeply cut leaves. Plants are covered by masses of fragrant blue-purple daisies, sometimes pink or white, all summer long.
◻ ✻

Calendula officinalis (Pot marigold)
A small, bright annual with orange daisy flowers in profusion from summer to autumn, especially in poor, well-drained soil. Self-seeds.
◻ ✻✻✻

Cosmos bipinnatus
Erect branching stems, up to 1.5m tall, bear delicate, cup-shaped flowers in shades of crimson, pink and white up until the first frosts. With its

ARGEMONE MEXICANA

soft, ferny leaves, it is a delightful plant for filling border gaps. The flowers of 'Sea Shells' have fluted petals. The shorter-growing Sonata Series is especially suitable for an exposed garden.
◻ ✻✻✻

Dimorphotheca pluvialis
An annual from southern Africa. White daisies, with blue-brown centres and blue undersides, appear above aromatic leaves. Flowers open only in sun. Plants reach about 40cm, and are good at the front of borders or in a container in a hot, sunny site.
◻ ✻

Eschscholzia californica ♀ (California poppy)
Bright, cup-shaped flowers, mainly yellow or orange, but also red, pink or cream, appear throughout summer above attractive, filigree, blue-green leaves. Plants grow to 30cm or taller. Long, curved seed pods follow the flowers. It self-seeds abundantly around the garden.
◻ ✻✻✻

COSMOS 'SONATA WHITE'

ESCHSCHOLZIA CALIFORNICA

LINARIA 'NORTHERN LIGHTS'

Lavatera trimestris
(Mallow)
Showy annual with lots of shallow pink or white trumpets all summer and softly hairy leaves. Grows to over a metre. 'Pink Beauty', with purple-veined, pale pink flowers, reaches only half that height, and glistening white 'Mont Blanc' is shorter still. (*See also L. 'Barnsley', p.54.*)
▫ ✱✱✱

Linaria maroccana
(Toadflax)
Erect plant, about 45cm tall, with small, snapdragon-like flowers in summer, mainly purple but also pink or white. Taller and long-flowering 'Northern Lights' includes shades of yellow, pink, salmon, carmine, lavender and white.
▫ ✱✱✱

Linum grandiflorum ♀
(Flowering flax)
Slender plants with narrow grey-green leaves are topped by loose spikes of rose-pink flowers. Grows to 40–75cm and is useful for softening stronger tones in a border.
▫ ✱✱✱

Onopordum acanthium
Extremely architectural, giant biennial thistle, forming a rosette of spiny, grey-green leaves in its first year, and 3m-high branching stems with purple or white flowers in its second summer.
▫ ✱✱✱

Papaver (Poppy)
Few annuals are easier, but be ruthless with self-seedlings. *P. rhoeas* Shirley Series, up to 90cm, comes in clear pinks, oranges and reds. At half that height, *P. commutatum* ♀ is red, blotched black. The opium poppy, *P. somniferum* ♀, about a metre tall, has bluish foliage and decorative seedheads.
▫ ✱✱✱

Portulaca grandiflora
A spreading annual, up to 20cm high, with red stems and fleshy green leaves. Single or double flowers, in pink, red, yellow or white, appear throughout the summer. The Sundance Hybrids have large, semi-double or double flowers in a wide range of colours. (*See also P. oleracea, p.33.*)
▫ ✱

PORTULACA GRANDIFLORA

SALVIA SCLAREA

Salvia sclarea (Clary)
A majestic biennial with eye-catching candelabras of pale pinkish-lilac flowers that shoot up in its second year, above large, wrinkled, highly aromatic leaves. Plants can reach over a metre tall. The variety *turkestanica* has pink stems and pink-flecked white flowers.
▫ ✱✱✱

Senecio cineraria
Although strictly a shrubby perennial, this is usually grown as an annual and a foliage plant. It forms 60cm mounds of felted, deeply lobed leaves of palest silver. Lower-growing 'Cirrus' is almost white – a striking contrast with other plants.
▫ ✱✱

Silene coeli-rosa
(Rose of heaven)
A slender plant with grey-green leaves and long-stalked clusters of dainty pink flowers with white centres, borne in profusion in summer. Grows to about 50cm. The flowers last well if cut for the house.
▫ ✱✱✱

INDEX

Page numbers in *italics* indicate illustrations.

ACKNOWLEDGMENTS

Picture research Sam Ruston
Illustrations Vanessa Luff
Additional illustrations Karen Cochrane
Index Hilary Bird

Dorling Kindersley would like to thank:
All staff at the RHS, in particular Susanne
Mitchell, Barbara Haynes and Karen Wilson
at Vincent Square; Candida Frith-Macdonald
for editorial assistance.

Photography
The publisher would also like to thank the
following for their kind permission to
reproduce their photographs:
(key: a=above, b=below, c=centre, l=left,
r=right, t=top)

Peter Anderson Photography: 35b, 35tr
Eric Crichton Photos: 39bl
The Garden Picture Library: Christi Carter
37tl, Andrea Jones 44bl, Ron Sutherland 5bc,
9tl, 11bl, 25cl
John Glover: 4bl, 9br, 10br

Jerry Harpur: front cover cla, 6, Marcus
Harpur: 42cr
Andrew Lawson: back cover tl, 28, designer
David Magson 44br
S & O Mathews Photography: back cover c,
8bl, 31bl
Clive Nichols: designer Ann Frith front
cover tl, Green Farm Plants /Piet Oudolf back
cover tr, Ann Frith 4br, 7bl, David
Stevens/Julian Dowle 23br, Green Farm
Plants/ Piet Oudolf 16
Howard Rice: 2, 46
Harry Smith Collection: 5br, 15tl, 24bl, 29bl,
38
Jo Whitworth: 24bc, 39br
Rob Whitworth: front cover bl, 14b, 27bl

The Royal Horticultural Society
To learn more about the work of the
Society, visit the RHS on the Internet at
www.rhs.org.uk. Information includes news
of events around the country, a horticultural
database, international plant registers, results
of plant trials and membership details.